The Environment and Science

Other Titles in
ABC-CLIO's

Science and Society

Series

Race, Racism, and Science, John P. Jackson, Jr. and Nadine M. Weidman

Women and Science, Suzanne Le-May Sheffield

Forthcoming Titles

Exploration and Science, Michael S. Reidy, Gary Kroll, and Erik M. Conway

Imperialism and Science, George N. Vlahakis, Isabel Maria Malaquias, Nathan M. Brooks, François Regourd, Feza Gunergun, and David Wright

Literature and Science, John H. Cartwright and Brian Baker

Advisory Editors

Paul Lawrence Farber and Sally Gregory Kohlstedt

The Environment and Science

Social Impact and Interaction

Christian C. Young

ABC-CLIO

Santa Barbara, California • Denver, Colorado • Oxford, England

Library of Congress Cataloging-in-Publication Data

Young, Christian C.
  The environment and science : social impact and interaction /
  Christian C. Young.
    p. cm. — (Science and society)
  Includes bibliographical references and index.
  ISBN 1-57607-963-5 (hardback : alk. paper) — ISBN 1-57607-964-3 (ebook)
I. Title. II. Series: Science and society (Santa Barbara, Calif.)

  GE50.Y68 2005
  304.2'8–dc22

                                                              2005002771

09 08 07 06 05   10 9 8 7 6 5 4 3 2 1

This book is also available on the World Wide Web as an eBook.
Visit abc-clio.com for details.

ABC-CLIO, Inc.
130 Cremona Drive, P.O. Box 1911
Santa Barbara, California 93116–1911

This book is printed on acid-free paper.
Manufactured in the United States of America

Contents

Series Editor's Preface

The discipline of the history of science emerged from the natural sciences with the founding of the journal *Isis* by George Sarton in 1912. Two and a half decades later in a lecture at Harvard, Sarton explained, "We shall not be able to understand our own science of to-day (I do not say to use it, but to understand it) if we do not succeed in penetrating its genesis and evolution." Historians of science, many of the first trained by Sarton and then by his students, study how science developed during the sixteenth and seventeenth centuries and how the evolution of the physical, biological, and social sciences over the past 350 years has been powerfully influenced by various social and intellectual contexts. Throughout the twentieth century the new field of the history of science grew with the establishment of dozens of new journals, graduate programs, and eventually the emergence of undergraduate majors in the history, philosophy, and sociology of science, technology, and medicine. Sarton's call to understand the origins and development of modern science has been answered by the development of not simply one discipline, but several.

Despite their successes in training scholars and professionalizing the field, historians of science have not been particularly successful in getting their work, especially their depictions of the interactions between science and society, into history textbooks. Pick up any U.S. history textbook and examine some of the topics that have been well explored by historians of science, such as scientific racism, the Scopes trial, nuclear weapons, eugenics, industrialization, or the relationship between science and technology. The depictions of these topics offered by the average history textbook have remained unchanged over the last fifty years, while the professional literature related to them that historians of science produce has made considerable revision to basic assumptions about each of these subjects.

The large and growing gap between what historians of science say about certain scientific and technological subjects and the portrayal of these subjects in most survey courses led us to organize the Science and Society series. Obviously, the rich body of literature that historians of science have amassed is not regularly consulted in the production of history texts or lectures. The authors and editors of this series seek to overcome this disparity by offering a synthetic, readable, and chronological history of the physical, social, and biological sciences as they developed within particular social, political, institutional, intellectual, and economic contexts over the past 350 years. Each volume stresses the reciprocal relationship between science and context; that is, while various circumstances and perspectives have influenced the evolution of the sciences, scientific disciplines have conversely influenced the contexts within which they developed. Volumes within this series each begin with a chronological narrative of the evolution of the natural and social sciences that focuses on the particular ways in which contexts influenced and were influenced by the development of scientific explanations and institutions. Spread throughout the narrative, readers will encounter short biographies of significant and iconic individuals whose work demonstrates the ways in which the scientific enterprise has been pursued by men and women throughout the last three centuries. Each chapter includes a bibliographic essay that discusses significant primary documents and secondary literature, describes competing historical narratives, and explains the historiographical development in the field. Following the historical narratives, each book contains a glossary, timeline, and most importantly a bibliography of primary source materials to encourage readers to come into direct contact with the people, the problems, and the claims that demonstrate how science and society influence one another. Our hope is that students and instructors will use the series to introduce themselves to the large and growing field of the history of science and begin the work of integrating the history of science into history classrooms and literature.

—*Mark A. Largent*

Prologue

When we read about environmental problems in the newspaper or on the internet, it is usually very easy to identify the human cause of these problems. Whether the latest concern is air pollution, dumping of raw sewage into important waterways, or rising global temperatures, the reports typically include a mix of scientific data and social concern. Within this mixture, there may also be hope that scientists will solve the problem. Put simply, humans create environmental problems with science and technology and hopefully can solve them with more science and technology.

If we admit that many of our current environmental concerns have a significant human cause, we can also be optimistic that recognizing the problem may lead to changes that will prevent a larger crisis in the future. In seeing the human cause and hoping for a human solution, we find the basic distinction between environmental science and environmentalism. Environmental science is the study of issues that affect the atmosphere, waterways, soils, plants, and animals. Such studies ultimately include humans. Environmentalism, on the other hand, is a social response to the findings of environmental science in combination with a wide range of economic, political, historical, and ethical perspectives. When it comes to global issues, like destruction of the rainforests or the melting of polar ice caps, scientists can often make clear statements about what is happening, although they do not always agree upon why. With those same issues, the broader response of environmentalists might involve a more complicated set of political, economic, and other agendas.

While scientists write reports, environmentalists stage protests. For example, each year since 1970, activists have designated April 22 as "Earth Day," and they work to raise awareness of environmental problems caused by humans. According to earthday.org, which lists events for every state, over sixty separate Earth Day events were scheduled in California alone in 2004. The events ranged from cleaning beaches and rivers to creating arts and crafts. Some events recognized the achievements of environmental activists over the previous year with awards; others provided demonstrations of earth-friendly products and technologies.

If environmental activism does not equate with environmental science, then educated citizens must begin to ask more questions about what role environmental science can play. How do environmentalists use the findings of environmental scientists? Are these scientists themselves environmentalists? Who provides funding for studies of the environment, especially when scientists discover trends that will be unpopular or require enormous sums of money to fix? When scientists themselves disagree, how do environmentalists know what to believe about the issues that concern us all?

The answers to these questions will involve more than a quick sampling of current issues. This book carefully reflects upon how environmental science has developed into a vast interdisciplinary endeavor. Few environmentalists realize how the movements they represent have been shaped by past events and concerns, and even fewer scientists are aware of the interesting history of their fields. However, this book is not just for environmentalists and scientists. Every student and citizen should be prepared to encounter today's environmental problems with a broader picture of how we got here.

Since every student and citizen has access to the environment and can make meaningful observations about their place in it, environmental science intersects with powerful personal and social beliefs. People have been making their own observations and developing ideas about their place in nature throughout human history. Therefore, modern science does not hold any monopoly on the subject of the environment. In this way, the history of environmental science must also include serious consideration of the many beliefs about nature that have persisted for centuries, as well as attitudes that emerge anew in each generation. Rather than taking on the enormous task of tracking these beliefs and attitudes, however, this book examines only a subset of those, where ideas about nature had a significant influence on developing scientific thought. In addition, environmental science has played its own role in shaping beliefs about and attitudes toward nature. This interaction between science and beliefs is examined in numerous places here.

Most people assume that nature is "out there," something scientists examine firsthand. In this view, industry can degrade nature, resource users can consume it, enthusiasts can appreciate and recreate in it, and the government can protect it. Under closer examination, this assumption seems like an oversimplification. Yet, when asking the questions "What is nature?" or "What is the environment?" we find a variety of new questions in our responses. Is nature a different thing to different people? How can this be, if nature is something "out there"? The answer may have something to do with how we each experience nature. Does the setting in which we were born and raised matter, such as a difference between urban and rural? Is nature different for people in different socioeconomic groups?

Many people also assume that scientists can be objective about the data they collect when they study nature. Because nature is "out there," scientists study to find out the truth about nature. While reasonably skeptical people might agree that scientists are not always objective, and most would recognize that scientists make mistakes, few students can imagine how a well-trained, diligent scientist could produce results that differ from another well-trained, diligent scientist. When their results do differ, students might initially wonder whether errors alone can account for the discrepancies. Specialized training might explain, in part, the different results, just as diverse educational and vocational traditions sometimes lead to dissimilar interpretations based on particular and distinct data. In other cases, individual scientists may prioritize their findings differently, depending on their perspective on the relationship between their results and the implications for broader questions for society. Only the most simplistic critique would view these different priorities as politically motivated, since the scientific process is a human endeavor in which individual decisions are always part of the procedure.

Despite the reality that the scientific process is always influenced by human decisions and interpretations, most people expect science to uncover "laws" of nature. Many of those laws (gravity and thermodynamics are two examples) prove to be extremely useful in our daily lives, although most people do not recognize their operation. Environmental scientists scrutinize a wide set of phenomena, and the usefulness of this information is often more subtle and difficult to recognize. These more remote phenomena interest us here, because it is important to examine the role of science in interpreting the data provided by nature. Students should be prepared to evaluate both the significance of data collected by scientists and the process of science responsible for collecting data. Although this may seem an overwhelming task, by looking at the content and process of science over the past two centuries, students can begin to see certain themes emerge that will help with their evaluation of more recent science.

Readers using this book may already be familiar with the broad outline of the history of western civilization. Some may also have some familiarity with environmental science, including ecology, meteorology, environmental chemistry, geology, and climatology. Some may even have taken courses in one or more of these rather specialized sciences. This book provides an introduction to the scientific concepts for those students who are new to environmental science. It is unlikely, however, that any students in the high schools and colleges of twenty-first century America have had the opportunity to consider the historical

A demonstrator holds up a mannequin's head wearing a gas mask during a rally in New York City on the first Earth Day. (JP Laffont/Sygma/Corbis)

background of these important sciences in a way that can enhance their perspective on the potential of science and enlighten their conception of the role of society in making crucial environmental decisions.

To underscore the significance of this history, we need only look as far as the reversals in American policy toward international cooperation in reducing global greenhouse gases. During the 1990s, the U.S. participated and provided leadership in establishing treaties aimed at protecting the environment on a global scale. While this participation seemed to threaten the progress of economic development in some industries, widespread support sustained policies that would limit air pollution and minimize atmospheric change. Since 2000, U.S. leadership and even participation in these efforts has evaporated. Despite overwhelming evidence from every field of science, changes in political leadership have led to dramatic changes in policy, largely in response to economic pressure. While these reversals are as disheartening to environmentalists as they are rousing to promoters of economic growth, changes in policy swing both directions. Since reversals occur regardless of scientific evidence, and sometimes in direct opposition to that evidence, there is no certainty that environmental policy will move in directions sought by environmental activists. Through historical examination of how environmental science has guided policy and responded to social needs in the past, we can learn to recognize the positive turns and minimize the negative swerves in policy and popular ideology.

The Environment and Science

Exploring a World of Possibilities

In the seventeenth and eighteenth centuries, explorers and colonists departed from their homes in western Europe. Beyond the familiar Mediterranean and European setting, they found a vast array of species and resources previously unknown to the scientific world. These discoveries and the subsequent settlement of new lands led to dramatic changes in their understanding of nature and their expectations for how nature could meet their needs. While some early explorers considered themselves naturalists, most of the men and women who left Europe in search of better lives in America faced their new surroundings without any specialized knowledge of nature. Advances in scientific knowledge came primarily from the European naturalists who searched for order in nature and endeavored to systematize their understanding. Explorers and naturalists established new systems in order to understand the environment and the role of humans in nature. Exploration marked the first step in the development of environmental science.

Even if they could not explore for themselves, wealthy men and women enjoyed both education and leisure time, allowing them to read about and discover the details of the natural world around them. The people who became most involved in the activities of observing and describing nature became known as *naturalists*. That term has taken on a wider range of meanings in more recent times, but for this period, natural historians examined plants, animals, and minerals. They developed collections of these objects and created various schemes to classify them. The common distinction "animal, vegetable, or mineral" comes from this phase of natural history. Other naturalists who concerned themselves more with theoretical and mathematical questions became known as *natural philosophers*. Many of their theoretical contributions remain significant today.

During the sixteenth and seventeenth centuries, natural historians and natural philosophers embarked on revolutionary studies of the world around them. They sought explanations for phenomena that had previously been viewed as miraculous, supernatural, or the work of the divine. In most cases, naturalists, as deeply religious men, believed they were studying God by studying nature. These naturalists hoped to illuminate the workings of nature as evidence of a divine creator. Far from seeing a battle between science and religion, these people

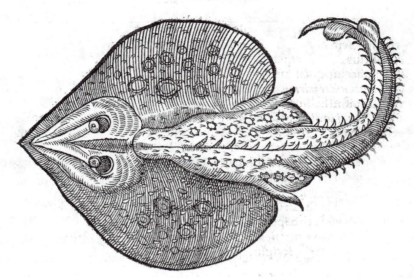

DE RAIA ASTERIA ASPERA,
Rondeletius.

Ray from Conrad Gesner's Historiae Animalium. *(Academy of Natural Sciences of Philadelphia/Corbis)*

enhanced their spiritual understanding of the creator by uncovering the details of the creation.

During this period, very few people had access to advanced education. Those who did generally studied theology and philosophy, combining natural history if their interests tended in that direction. Generally, only men found opportunities for education, since careers in the church or society did not exist for women and it hardly seemed appropriate that rare educational opportunities should be provided to them. Nevertheless, some women managed to play an active role in the natural history of their day. They contributed as illustrators and recorders, making observations and offering explanations that typically became the intellectual property of a man, who could publish the work. Often these women worked with their husbands, brothers, or fathers, receiving little recognition but enjoying for themselves the intellectual rewards of the work.

Naturalists like Conrad Gesner (1516–1565), Andrea Cesalpino (1519–1603), and John Ray (1627–1705) conducted broad surveys of their local flora and fauna, compared their collections with specimens from the far reaches of the newly explored globe, and compiled their findings in massive, illustrated texts. In the middle years of the sixteenth century, Gesner collected specimens from across Europe and the newly explored world beyond, although he lived most of his life in the cities of mountainous Switzerland. He became one of the first naturalists to choose not to rely upon the ancient philosophers for accurate accounts of nature. He studied nature as it appeared in diverse places in his own era.

Gesner believed that the work of ancient philosophers offered little guidance in identifying the broad range of plants and animals he encountered on his travels and in collections across Europe. Although he admired the ancient texts, he sought to correct them with respect to his own collections, which were as vast as any of his time. Over the course of almost a decade, he wrote an encyclopedia entitled *The History of Animals* to update ancient descriptions. Gesner also hoped to explain the relationships of animals in their natural settings. Since his first task was identification, he did not make much progress in developing the framework of animal relationships, but the need for such a plan opened new possibilities in the literature of natural history in the scientific revolution.

Around the same time as Gesner, Italian naturalist Andrea Cesalpino concentrated his efforts on plants and hoped to devise a classification system based solely on their biological characteristics. Cesalpino was inspired by Aristotle to identify the essence of each species in order to determine its place, rather than rely on usefulness to humans as a criterion as earlier authors had done. This return to an ancient principle made it possible for Cesalpino to study plants in a way that separated them from the functions assigned to them by medical doctors and herbalists. As a result, he used nutrition and reproduction as independent functions of each species to group them into broader categories. When later naturalists established new categories, rejecting those of Cesalpino, they nevertheless retained his basic system of classification.

In the seventeenth century, John Ray, an Englishman, acknowledged the revolutionary work of Gesner, Cesalpino, and others. Ray believed that natural history provided a powerful means of connecting beliefs about God and the creation of the world around him. Rather than take revolutionary approaches to nature as a step toward mechanical explanations or atheistic descriptions, he found new insights into the natural history of the scientific revolution. He committed himself to the idea that nature demonstrated the works of the Christian God in his *Wisdom of God Manifested in the Works of the Creation.* Like other naturalists of his day, Ray believed that the conceptual classification schemes developed by naturalists were a reflection of a divine creator's plan. The imposition of a belief in God served as an inspiration to natural history rather than an obstacle. He also offered examples of the many ways people could benefit from a better understanding of nature. He believed understanding would lead to progress in human society by recognizing the opportunities and resources made available by a wise God.

Ray did not contribute as much to the knowledge base of natural history as had Gesner or Cesalpino, but his unique articulation of the implications for that knowledge provided a distinctly Christian and Protestant answer to the mecha-

Conrad Gesner (1516–1565)

Conrad Gesner was born on March 26, 1516, in Zürich, the son of a poor fur dealer and his wife, Ursus and Agatha (Frick) Gesner. He learned something about animals from

his father, and in the garden of his great-uncle he began to learn about botany. His great-uncle raised him when his parents could not afford the expense of another child in their large family. Gesner was sent to the German elementary school in Zürich, but the teachers soon recognized that he had a gift for languages, and he went to a more advanced school to study classical languages. To continue his education, Gesner considered studying theology, then turned to medicine instead. Before long, he changed his mind again, preferring the challenges of studying classical languages.

Gesner married Barbara Singerin in 1536. The only job he could find in Zürich at the time was that of a low-paying grammar school teacher. In his spare time, he read the works of the ancient physicians, reentered medical school, and studied classical languages. He also began work on

Naturalist Conrad Gesner (1516–1565) published descriptions and drawings of plants and animals, both real and fanciful. (National Library of Medicine)

his first major publications: *Historia plantarum* and *Catalogus plantarum*, which appeared in 1541 and 1542. These works included a large number of plants listed

nistic views of philosophers, especially in Enlightenment France. He opposed philosophical ideas that saw nature as a machine. This mechanistic philosophy often accompanied a denial of God's role in nature, and even the very existence of a divine being. For Ray and others, the mechanistic philosophy provided a new incentive to examine nature carefully in search of evidence of God's goodness. Ray provided a text for seventeenth century English naturalists and theologians that held the Protestant worldview together in the face of Newtonian innovations. Those innovations, which many saw as a departure toward a mechanistic view of nature, could and would be interpreted in many ways. Ray maintained a place for God in creation.

The System of Nature

Born Carl von Linné (1707–1778), the Swedish naturalist began referring to himself as Carolus Linnaeus when he undertook the project of naming every plant, alphabetically for easy reference, and included medicinal uses of the plants. While writing them, Gesner completed medical school.

On the rare occasions when Gesner left Zürich, he invariably spent his time traveling to places of botanical or zoological interest. In the years from 1551 to 1558, he compiled the most extraordinary work on animals ever attempted, his *Historia animalium*, which consisted of four volumes and contained over 1,500 illustrations. Each volume dealt with a separate category of animals, approximately following Aristotle's scheme: the four-footed animals that gave birth to living young, the four-footed animals that laid eggs, the birds, and the fish.

His travels to collect plants took him frequently into the Alps, and he gained a reputation as one of the earliest mountaineers. Throughout the Middle Ages, people regarded the mountains with dread and often outright fear. Gesner helped to overcome this with some of the earliest known writings praising the beauty of the mountains. A letter to a friend in 1541 has often been quoted for its uniqueness in literature at the time and also as an inspiration to modern mountain climbers.

Medicine was a unifying principle in Gesner's study of nature. His interest in ancient works stimulated his deepening study of plants and animals during the first half of his life, but later, medicine became the driving force of his studies. As plagues raged across Europe, he became more and more determined to understand the diversity and secrets of nature in the hope of finding cures and remedies. In the process of scouring nature for these answers, he compiled a great deal of knowledge. When Conrad Gesner died peacefully in his own museum on December 13, 1565, he left a great deal of work undone.

animal, and mineral with a standard form in the classical scientific language of Latin. As a young naturalist, Linnaeus first relied on his own experience exploring the Arctic. He studied species in Lapland, in the northern reaches of Scandinavia, in 1732. This five-month excursion above the Arctic Circle became an important period in his development. During his travels, Linnaeus discovered the arduous realities of field botany, collecting plant specimens and recording his observations. When patrons later offered him the opportunity to travel to Africa for more collecting, he chose to stay in Europe amid the comforts of a familiar community and specimens brought back by others.

Linnaeus encountered an even larger challenge than his sixteenth- and seventeenth-century predecessors, who had updated, translated, and interpreted many of the surviving ancient works. In the course of traveling around the globe, explorers and merchants had returned from Asia, the Pacific, and the Americas with various new plant species. European naturalists struggled to determine the place of the staggering new array of species in their existing categories and systems of classification. In addition to the enormous numbers of new species, it

Carolus Linnaeus (1707–1778) was a Swedish botanist and taxonomist, considered the founder of the binomial system of nomenclature and the originator of modern scientific classification of plants and animals. (Corbis)

came to be recognized that many of these plants grew in climates similar to the familiar European climate. Yet, despite the similarity in climate, the plants exhibited tremendous variation when compared with the familiar species of Europe. Fitting these new forms into existing classification systems presented a mounting problem to botanists, which Linnaeus recognized. Another complexity that these new species introduced was the familiarity of plants, which seemed similar to European forms, but which originated across vast oceans in distant lands. These complexities of species variation pressed European naturalists to reconsider their categories and reimagine the source of all this diversity.

Linnaeus stood convinced that the great complexity of plant life provided further evidence of God's inventiveness. He decided to devote himself to research in botany. He hoped to establish, if possible, a new classification system that would reflect God's design in nature. Linnaeus believed that, however complex it seemed, the system would ultimately reveal an orderly plan rather than a haphazard array of species scattered indiscriminately around the globe.

One of the keys to establishing an orderly classification system, Linnaeus believed, was to implement a standard system of naming plants, animals, and minerals from foreign lands. Linnaeus knew that local people in different regions had distinct names for the species they knew. When travelers brought specimens from the Americas or Asia, likewise, they often called the species by the local names used by the people in those places. For Linnaeus, these names were not useful and were even misleading at times. Names for different species from different places might be similar, but the names coincided by chance rather than by systematic efforts to group them. He proposed a new naming system that would supercede all names from local areas. Linnaeus hoped his naming system would simplify the confusion that resulted when a single species had multiple names in different areas or when two or more different species shared the same name. It offered the attraction of standardization. The naming of each species would follow a form borrowed by Linnaeus from previous naturalists but, for the first time, implemented on a universal scale. He insisted that each species be given a two-word name, one that placed it in a category of closely related species and one that identified each species as unique within its category.

Linnaeus's system has endured many changes and additions in biological thinking. One might suppose, therefore, that he succeeded in creating a scheme that reflected and even encompassed the "real" plan of nature. Such an explanation for success would have surprised even Linnaeus. Another explanation for this success focuses on the relative simplicity of the logic he followed in devising his system.

A model for the logical system Linnaeus used appeared over a century earlier in a book entitled *Isogogethica ad rationis normam delineata* (An introduction to ethics as delineated by the rule of reason) by Nicholas Abraham. In that book, Abraham reduced the decision-making process to a set of choices. In each choice, there were only two options. By establishing the choices in sets of two, he followed a system dating back to Aristotle, where each choice creates a dichotomy. For example, when one approached an ethical decision, the first choice was to opt for a course of action that would follow one's own way of thinking or to follow the custom of one's culture. In the case of Linnaeus's classification system, he examined the characters of plants and animals, deciding which characters could provide simple choices in arranging them into groups. Generally, the first choice was to decide whether an object was a plant or an animal. Then, the classification of a familiar animal would involve choices based on some of the following questions: Does it have a backbone? Does it give birth to live young that are then nursed by their female parents? Does it primarily eat meat? If yes to each of these, it could be a species in the order of carnivores, which belongs within the class of mammals. One would have to go farther to determine whether the species in question belongs to the genus Canis (the dogs), Felis (the cats), or another genus. If it were a dog, it might be the coyote (*Canis latrans*), the wolf (*Canis lupus*), the domestic dog (*Canis familiaris*), or another species of dog.

The system of logic followed by Linnaeus proved extremely valuable in grouping plants and animals into meaningful categories. He imagined categories as being quite rigid in this system. Linnaeus thought of them as so many numbered pigeon-holes that could be filled as naturalists discovered and named new species. In the end, there would be no gaps between species, since Linnaeus believed the creator had made a complete grid without open spaces. Since he believed that order reflected the design of a creator, he continued to work and refine his system throughout his life. The classification became somewhat more flexible as explorers discovered more new species, and Linnaeus eventually compromised a bit on his insistence that species could only fill places already existing in the scheme. Explorers and amateur naturalists who discovered what they hoped were new species acknowledged Linnaeus as the authority on such discoveries and took pride in having their species listed in his work. After the first version, published in 1735, Linnaeus updated and expanded his *Systema naturae* in subsequent editions. He laid out his naming method in the tenth edition in 1758. The revisions led to even greater use of his work. A British committee adopted the twelfth edition of *Systema naturae* as their standard reference in the nineteenth century. The Linnaean system spread across Europe at different rates. In Paris, the French accepted Linnaeus's list of plant names in *Systema plantarum* as their standard in 1867. By the nineteenth century, the *Species plantarum*, published in 1753, served as a standard reference book for classifying species from around the world. He based it on his collection of plants, which eventually consisted of 19,000 pages of pressed specimens. His collection included 3,200 insects and 2,500 minerals, which he also attempted to include in his system of nature.

The remarkable structure Linnaeus brought to classifying natural species had one limitation: it did not work for minerals. Although he tried throughout his life to find ways of applying his general system of categorizing plants and animals to the third great division of natural history, a consistent arrangement for minerals eluded him. Neither the logic he employed nor his belief that he might ultimately uncover the creator's plan could overcome the complexities of mineral types and forms. Particularly perplexing were those "species" that resembled so closely the forms of living things. Commonly known as fossils, there was no explanation for the resemblance. Explorers had collected an astonishing array of fossil forms with similarities to living species. Naturalists generally agreed that certain forms were common in nature, and that both living and nonliving things, by some unknown mechanism, would develop into those forms. During his lifetime, the alternative explanation provided by George Buffon offered Linnaeus no solution to the problem of trying to classify *all* minerals according to his own broader system.

Linnaeus nevertheless received international acclaim for making sense of diverse living forms. His system encompassed known species as well as the increasing numbers of newly discovered species carried back to Europe by explorers. Linnaeus's contribution took on enormous significance since new expeditions embarked constantly in search of more efficient trade routes and undiscovered resources. The search for order could have easily overwhelmed a less comprehensive system.

The Order of Nature

While Linnaeus hoped to clarify the place of each species of plant, animal, and mineral in God's design, the director of the Royal Garden in Paris took an opposite approach. George Louis Leclerc comte de Buffon lacked the training in natural history that most botanists and even physicians received at that time. When Buffon took over the botanical and natural history collections gathered from around the globe, he took as his first responsibility the task of cataloging the specimens.

For Buffon, the interesting challenge at the Royal Garden involved uncovering the order of nature. Unlike Linnaeus, he did not assume any divine plan. Rather, he hoped that with such a vast collection at his disposal, he could create the system through reason alone. Buffon represented the approach to nature that John Ray had opposed. This mechanistic approach fit neatly into the broader culture of French thinking at the time, known as the Enlightenment. Buffon's writing ultimately became the most significant description of nature in Enlightenment literature.

It seems somewhat surprising to find Linnaeus and Buffon differing so markedly in their approach to studies of the natural world, living at the same time as they did. However, while their motivations differed for reasons of their divergent training and the societies in which they lived, they shared many commonalities. Both enjoyed the patronage of extremely wealthy benefactors, although they each lived on comparatively modest salaries. They both had access to enormous collections upon which they were free to conduct comparisons and arrange categories. Both men corresponded with a vast and overlapping network of travelers, explorers, merchants, and naturalists who shipped a constant supply of new specimens from all over the world back to Europe. Certainly not least of their similarities was their shared respect for the awesome complexity of nature and its intricate beauty.

Since Buffon began his search for the order of nature without a preordained notion of how the world was created, he looked for causes and connections between species that theological naturalists had taken for granted as divinely created. He recognized through his studies and comparisons the link between nature in the present world and nature in the past. Living plants and animals possessed similarities to forms found preserved in layers of rock that had been laid down in ages past. These similarities were familiar to Linnaeus as well, although he had difficulty accounting for them. Naturalists called these rocks "fossils" and had wondered since ancient times how they came to resemble living beings. Hardly satisfied to imagine that sheer coincidence could create such resemblance, some naturalists theorized that certain forms were natural and as likely to occur in minerals as in plants or animals. Invoking the divine plan of a creator reinforced these theories; God could create as many or as few forms as He liked and manifest them in any material substance.

Buffon, who became curious about the potential historical links in nature and did not concern himself with supposed divine plans, proposed a different explanation. In fact, his explanation is so widely accepted now that it is difficult to imagine the notion that similar forms arose independently in living and nonliving materials. Buffon joined those who suggested that the forms so similar to living beings found in rock layers were the direct remains of impressions made by plants and animals themselves. In some cases, minerals deposited over time within the tissues of plants and animals after they died retained the form of the original, creating solid replicas. While those tissues decayed relatively rapidly, the impression of the living being remained in the mineral or sedimentary rock layer.

Connecting the natural world of the present to that of the past yielded other insights for Buffon. When confronted with the remains of species apparently no longer in existence, other naturalists made the assumption that those species had gone extinct due to some unappreciated whim of God's will, or that those species still existed in some as yet undiscovered corner of the world. Buffon offered another alternative. He suggested that nature was completely full, and that new species could both come into existence and go extinct from time to time. The total number of species in nature would thus remain constant. It was no judgment on any species. Instead, it suggested to Buffon that the order of nature defied simple explanations. He was thus motivated to search more diligently for the sources and components of that complex order.

The Age of the Earth

During his lifetime, Buffon began an encyclopedia entitled *Histoire naturelle*, which took his assistants twenty years after his death to complete. *Histoire naturelle* filled thirty-six volumes and included information in the form of text and illustrations that surpassed any previous collection. Within this work, in exploring the order of nature, Buffon proposed an explanation for the origin of the earth itself. His theory included components of physical and mechanical explanations compatible with Newtonian physics. A comet colliding with the sun, Buffon proposed, might throw enormous amounts of matter into space. Over a period of perhaps 70,000 years, that glowing-hot molten material from the sun could have cooled and solidified into planets. One of those planets, Earth, had mountains of that original material as well as other deposits of matter from later periods. Buffon also proposed that, as the Earth cooled, the condensation and falling of water as rain from the superheated atmosphere formed the oceans. Species that lived in the early oceans were adapted to the high temperatures of this young earth. As the earth continued to cool, those species went extinct and were replaced by the species known to the naturalists of Buffon's day. This explanation helped to account for the forms of plants and animals found by explorers that had no living counterparts. The same sequence of adaptation to a warmer climate, extinction, and replacement by new species was repeated on land. Buffon thus provided a coherent account of earth history that was generally consistent with the laws of physics understood at that time.

Geological debates particularly provided a starting point for making sense of the tremendous amount of information coming from explorers from around the world as they brought back descriptions of landforms and samples of minerals and fossils in the late eighteenth century. Until that time, most naturalists recognized the creation account provided in the Bible as a sufficient explanation for the origin and placement of the oceans and solid ground. They called this explanation "Bible Geology." Genesis contained a record of how God had separated day from night, the heavens from the earth, and the waters from the dry land in the first days of creation. God completed the process by the end of the sixth day by creating living things, the sun, the moon, and the first humans. Genesis offered no details for how God accomplished all of this, and naturalists had collected material for centuries in the hope of coming closer to an understanding of God without necessarily demanding or even expecting to know how creation was accomplished.

Part of the explanation that many naturalists accepted involved the story of a worldwide flood, from which God had spared only Noah and his family. The flood destroyed all living beings on the earth except those saved aboard Noah's ark, which God had instructed him to build. When plants, animals, minerals, or fossils appeared in unexpected places, naturalists often ascribed the occurrence to this flood. Many cultures possessed narratives of some massive flood that had virtually destroyed humanity, so while most naturalists shared the Judeo-Christian account of Noah, they could point to other cultures in support of this universally recognized natural phenomenon. A worldwide flood explained certain features of erosion and the catastrophic disappearance of many life forms. Genesis provided sufficient details of the Earth's origin and catastrophic changes to satisfy most reasonable people

By the end of the eighteenth century, however, some reasonable people had seen evidence of a much older Earth, and the possibility of a worldwide flood was contradicted by geological evidence from around the world. Traces of marine life on mountaintops seemed the improbable result of a forty-day flood. Instead, geologists suggested that the floor of the ocean might have been pushed upward over eons of time, eventually carrying sediments loaded with seashells to the slopes of mountains. In other places, landforms appeared to result from millions of years of gradual, geologic change that disturbed underlying layers only slightly, while a catastrophic flood would have mixed and dispersed those layers. Time had preserved a record of the Earth's age that seemed incompatible with Bible Geology.

Some naturalists, like Buffon, adopted an increasingly mechanistic worldview, with mathematical formulations championed by Isaac Newton and others. Many natural historians and philosophers searched the material world for theories sufficient to explain the workings of nature through mechanical principles and mathematics, without the intervention of a divine creator. Most of these naturalists accepted some notion of a creator who set the universe and the laws of nature in motion, but many began to consign this creator to the role of a clockmaker, now sitting back and watching the device run on its own. Naturalists who looked to the fluid force of water, through processes including flooding and erosion, were known as neptunists. Their views contrasted with those of the plutonists, naturalists who saw geological forces from volcanoes and the internal heat of the earth as constructive and destructive. In either case, a creator took a secondary role. Combining the evidence of an old earth with this mechanistic worldview provided a new perspective from which to construct geologic principles.

The English geologist Charles Lyell (1797–1875) became the first to propose an explicit alternative to Bible Geology in a book titled *The Principles of Geology* (1830). In this three-volume work, Lyell demonstrated his view that all of the geological features now visible on the Earth were the result of processes that were currently under way and ongoing. Both neptunists and plutonists provided Lyell with valuable insights. Each explanation offered reasonable contributions to the idea that the Earth was old and that it had changed over time. Earthquakes, volcanoes, and erosion could each explain local and relatively small-scale changes on the Earth's surface. Although human communities living in the presence of these changes sometimes considered them catastrophic, observable processes could account for high mountains, deep gorges, and the deposition of sediments across large landscapes.

Lyell therefore rejected the idea of worldwide catastrophic processes to account for the formation of such features. Because the rate of change in nature described by Lyell in *Principles of Geology* was constant, or uniform, and the processes were ongoing, this view became known to naturalists as *uniformitarianism*. Because the changes were gradual rather than abrupt, the view was also known generally as *gradualism*. The opposing viewpoint, *catastrophism*, assumed that worldwide events with unknown causes, such as the six-day creation or Noah's flood, explained the earth's present form. If the general public could still contentedly believe in the Genesis account of creation, then those taking the study of geology seriously began to move beyond popular belief. Lyell worked to uncover both the process and the sequence of geological time.

Alongside Lyell, several British and French naturalists established the sequence of rock layers, called strata. They found this to be an exceedingly complex task. The French natural historian Georges Cuvier (1769–1832) explained gaps and abrupt changes in the layers as evidence of catastrophic events in Earth history. In England, the scientific community was embroiled in a debate over how to classify the strata. They worked from samples collected mostly on the surface, although projects like railroad construction blasted away parts of hills, directly exposing the layers. Putting the strata in order assumed that the most recent sediments would be found on top of successively older layers. It became known as the principle of superposition, and it seemed simple enough. Upper layers were equivalent to younger layers; lower layers were older. However, in some areas more recent sediments might be eroded away, leaving older layers on the surface. Much older strata might also be deformed by movements even deeper within the Earth, which could twist, crack, and even lift the older layers to the surface and turn them on top of newer layers. Within many of these layers, fossils provided evidence of the living organisms that existed during the period of that stratum's formation. By comparing fossil forms from relatively stable strata, naturalists could reconstruct many of the changes that took place in areas with more upheaval. By these reconstructions, they could determine when deformations and subsequent erosion had displaced the expected sequence. Eventually, the great controversies gave way to an agreed upon sequence of geological "formations" describing the major periods of Earth history.

The adoption of a standard sequence was not the only concern of geologists in the nineteenth century. Like other naturalists, geologists sought a connection between the processes of a changing Earth and the living species that inhabited it. If the Earth were gradually and continually cooling, as Buffon had suggested, then perhaps species went extinct if they had developed adaptations to warm weather but had lived near the poles where the climate had since become quite cool. Many fossils of plants considered to be of tropical origin existed in temperate regions far north or south of the tropics. Animal distribution exhibited the same pattern. Explorers found remains of species similar to elephants near the Arctic and wondered whether the climate there might once have been warm like central Africa. Naturalists eventually realized that these "wooly" mammoths and mastodons probably survived in the enduringly cold region where they were found, but it took some time to answer this question definitively. If geological change played a part in the development of living species, naturalists hoped to identify the connections.

Lyell explained that much of the evidence for a cooling Earth actually supported his theory of uniform change. While his opponents used catastrophic events on the surface of a planet that had been significantly warmer to explain its features, Lyell pressed his uniformitarian views, arguing that gradual warming and cooling had cycled throughout Earth's history. Fossils of tropical plants in temperate regions suggested to Lyell that some areas were warmer during certain periods in the past, but not that the entire planet was once much hotter and had regularly cooled off since.

Lyell's uniformitarian geology was inspired in part by the study of volcanoes, which had grown to include examples visited by explorers around the globe. Upon visiting Italy's famous Mount Etna, he saw for himself that the massive mountain had been laid down over a long period of time through multiple eruptions. Layer upon layer of lava accumulated to form the giant cone of the volcano. Compared with the formations of other regions, the underlying rock of the area surrounding Mount Etna was relatively young. For Lyell, the fact that human observers saw the mountain as very old reinforced his view that such a perspective would not suffice for a geologist. The age of the volcano must also be reconciled with the relative youth of the formation on which it stood. Just as a house could not be older than its foundation, no volcano could be older than the strata below it.

The process of reasoning Lyell followed allowed him to imagine changes over vast periods of time. These changes, such as the elevation of land caused by a minor earthquake, could accumulate over time. If one earthquake, he reasoned, could raise a shelf of land just five feet along a stretch of many miles, the accumulated effects of 2,000 earthquakes over eons of time would build a mountain range 10,000 feet high. He surmised that just such a process had occurred along the coast of Chile in South America, where such earthquakes occurred about once every hundred years. As added support for this explanation, which took vast periods of time, the changes would be gradual enough to allow plant and animal life to persist in the area with minimal disruption. If an entire mountain range rose in just a century, the consequent disruption to living things would mean extinction for a much higher proportion of species than naturalists had observed in such areas. Changes in ocean levels and the coverage of land masses by ice sheets could similarly be understood as gradual and ongoing geological processes.

While Lyell's approach involved a certain amount of speculation about the processes that led to the current state of nature, he insisted that all of the processes currently functioned and were observable in nature. He did not include supernatural intervention or other drastic departures from the laws of nature. Instead, Lyell restricted his explanations to the regular workings of volcanism, earthquakes, and erosion to describe the building up and tearing down of landforms. Such a restriction also allowed him to use the rate of previous events to achieve a level of predictive power not available to catastrophists. By calculating previous rates of erosion, distance, and depth, Lyell could predict how long it would take for Niagara Falls to erode its way to Lake Erie, which lay twenty-five miles to the south. Since the falls had moved an observable fifty yards in the forty years prior to his writing, and with another twenty-five miles to go, Lyell predicted that if current rates persisted, the falls would reach the lake in another 30,000 years, or slightly more. Lyell could therefore predict a massive flood, 30,000 years hence, resulting from the rapid draining of Lake Erie into the flat lands around Lake Ontario.

Charles Darwin (1809–1882)

Charles Darwin's most famous book, *On the Origin of Species*, was the product of decades of field research and theoretical work. He published it in 1859, when it

became clear that another man, Alfred Russell Wallace, had proposed a mechanism for evolutionary change that resembled Darwin's so closely that the two would likely share credit for the idea. In fact, they did share whatever glory there was within the scientific community, but Darwin had articulated the idea sooner and elaborated his thesis in much greater detail, so that the general public recognized him as the primary author of the evolutionary mechanism of natural selection.

Darwin was born on February 12, 1809, the same day that Abraham Lincoln was born. Just as Lincoln never imagined as a young man that he would someday be president of the United States, Darwin could never have envisioned the fame that awaited him in a scientific career. Although his father hoped Charles would study medicine and become a physician, the younger Darwin was repulsed by his early experiences in

Charles Darwin (1809–1882) formulated and popularized the mechanism of natural selection to support the controversial theory of evolution in the mid-19th century and published On the Origin of Species *in 1859. (Library of Congress)*

medical school. He eventually studied to be a clergyman, while his true ambition at that point was to study natural history. A naturalist could scarcely make a living in

Before that happened, Lyell noted that the size of Lake Erie might have contracted due to sedimentation. The contraction might be enough to make the predicted flood rather insignificant. Given the number of qualifications he offered in this explanation, Lyell clearly was not in a position to predict accurately the outcome of thousands of years of geological change in the future. He could, however, make reasonable estimates of the range of possibilities, without introducing supernatural intervention, which would presumably rely on the whim of a divine power and could not be predicted. Scientists who observed further changes in subsequent years could improve on Lyell's predictions and correct assumptions that proved wrongly based, but they would have a consistent framework on which to base their work.

The explanatory power of the uniformitarian approach led some natural historians to develop explanations for phenomena that previously eluded them. Naturalists struggled to explain, for example, the existence of large beds of rock and huge boulders on dry land similar to what they would expect to find at the

nineteenth-century England, but a country minister might hold a respectable place in society and still have time for the pursuit of botany, entomology, and the like.

While Darwin studied at Cambridge, he enjoyed his natural history courses most of all, and through one of his professors received an endorsement to accompany a ship's captain on a voyage to South America aboard the *H.M.S. Beagle*. There, the expedition was to complete a mapping project and collect specimens of plants and animals for further study back in England. Darwin proved to be so skilled as a collector that he took over the job of ship's naturalist and persuaded the captain to continue the voyage around the world. By the time they got back to England, the expedition had been away for five years. Darwin had earned a fine reputation as a collector and the right to correspond with the finest naturalists in the world. He spent many years examining and classifying the specimens he had collected. One of the questions that plagued him during these years was how to account for the tremendous diversity of species scientifically without recourse to supernatural explanations.

After working on the *Origin* and several other projects, including studies of coral reefs and barnacles, Darwin found himself among the most sought after scientists of his century. He continued to struggle with the larger philosophical and theological questions raised by evolution, although he kept out of public debates on the subject entirely. In 1871, he published some of his conclusions about the place of humans in evolutionary history in *The Descent of Man*. He wrote comprehensive treatises on orchids, insects, earthworms, plants that climb, and plants that eat insects. Darwin's extensive skill as a naturalist would have earned him a place of honor among his colleagues even without his contributions to evolutionary theory. He is remembered as one of the greatest scientists of all time.

bottom of a slow-moving river. Catastrophists could fall back on the idea of a global flood that had deposited gravel in unexpected places. Uniformitarianists did not have a ready answer. By the 1830s, Jean de Charpentier (1786–1855) and Louis Agassiz (1807–1873) independently proposed glaciers—massive ice sheets—as an alternative to a great flood. Agassiz's version of glaciers covering large sections of the northern continents received more attention than Charpentier's, perhaps in part because Agassiz made his ice ages resemble the catastrophes that were still well accepted by most geologists. Charpentier proposed glaciation as a gradual process resulting from long periods of cooler weather, not a major climatic change. Ironically, Agassiz was himself a catastrophist, and consequently Lyell and other uniformitarianists at the time rejected his explanation. Because of their differences and Agassiz's well known preference for catastrophe, studies of glaciers did not become integrated into uniformitarian geology until late in the nineteenth century, and causal accounts of ice ages were unknown until even later. Nevertheless, naturalists increasingly recognized the likely role of ice in causing erosion and depositing sediments in ways and places that water alone could not do. Charles Darwin was among the earliest to embrace ice ages as an explanation for change in the natural world, even while Lyell continued to reject them.

Darwin's Search for Explanations

Perhaps Lyell's greatest contribution was his search for mechanisms that could account for observed processes and forms. He was inspired by the mechanists of the previous century, and particularly by Newton who had postulated forces in nature governed by laws that could be discerned mathematically. Charles Darwin drew inspiration from the same tradition, and was especially influenced by Lyell's geological works.

Like many naturalists of the early nineteenth century, Darwin hoped to make his reputation as a scientist by contributing the kind of explanation that Newton had offered. Darwin's love of the natural history of living things provided an arena in which he might make such a contribution. As a student, he read William Paley's *Natural Theology* (1802), which argued that organisms functioned perfectly because they were specially designed for those functions. Darwin found Paley's argument elegant and highly persuasive, and he looked forward to examining the evidence of nature himself. In planning for a possible career in medicine, Darwin realized that he would never succeed as a physician, much to the disappointment of his father. He hoped ultimately to become a reputable naturalist, but recognized that in order to maintain the family's place in society, he would first have to establish himself in some other profession, if not as a doctor then perhaps as a clergyman. Such a life would afford him opportunities to expand his interest in natural theology.

While preparing for a profession in the clergy at Cambridge, Darwin studied with a well known botanist, John Henslow. This experience set in motion his life's work, for it was Henslow who helped Darwin gain passage on a surveying expedition around the world aboard the *H.M.S. Beagle*. Henslow also gave Darwin a copy of the first volume of Lyell's *Principles of Geology* to read on the voyage. By the time Darwin left on the *Beagle* in 1831, he had met some of the leading scientists in Britain and his interests were known to many. A trip to South America and beyond could have meant isolation from the society with whom Darwin had so recently become acquainted. Instead, Darwin actually increased his visibility among those scientists by regularly shipping specimens back to England and by corresponding with Henslow and others.

Darwin's extraordinary skill in describing and cataloging plant, animal, and fossil specimens as he traveled around South America, to the Galapagos Islands, and eventually across the Pacific, emerged as even more significant in his development as a well regarded naturalist. The process of collecting specimens constituted only the first step, for if specimens were not also properly preserved for storage and transportation, collection efforts would be wasted. Darwin personally attended to every step, from collecting to preserving, and for each specimen he wrote a careful description. He accompanied many of his descriptions with sketches. He filled his notebooks with details about where each specimen was found, as well as the conditions of weather and geography. For fossils, he added questions that came to mind about the position of each fossil in relation to layers of sediments and why the species had gone extinct. At every port, Darwin enjoyed his time away from the ship, collecting and exploring, often on his own. Onboard the *Beagle*, he remained almost constantly seasick, but he managed to catalog his growing collection, so that when he returned to England he had recorded virtually every detail for further comparison and study.

Darwin read constantly during the voyage. In addition to Lyell's book, he reread the great explorers' accounts of his day, including Alexander von Humboldt's *Personal Narrative*, a seven-volume book loaded with details about Humboldt's own trip to South America some thirty years prior. Darwin's inspiration to travel had come in part from his first reading of this book, and during his voyage in the 1830s, he studied it carefully, comparing Humboldt's experiences to his own. One of Humboldt's essays, "Geography of Plants," questioned the placement of different species around the world in surprisingly consistent ways. Certain patterns, complex but perhaps ultimately comprehensible, begged for an explanation. Humboldt searched for an integrated vision of nature, but he fell short, even in 3,754 pages. He devoted many of those pages to sweeping descriptions of what he saw. For all the wonder that Humboldt had inspired, Darwin realized that the words of description still could do no justice to the inexplicable beauty he found in South America.

The writing Darwin did onboard the *Beagle* eventually rivaled Humboldt's, but the English naturalist eventually came back with something even more significant. Darwin was undoubtedly inspired by Humboldt's vision of integrating nature, and yet he hesitated to draw sweeping conclusions. He considered himself a natural historian of very limited reputation. Among the general public, readers might appreciate another account of travels in the tropics, but Darwin had become acquainted with a circle of naturalists in London by whom he wanted to be taken seriously. He wanted to produce a work of even greater significance than had Humboldt.

Upon his return to England in 1836, Darwin's reputation as a naturalist was more secure than he might have imagined. He had kept up correspondence with many of the leading scientists of society he met before he left, including Henslow and Lyell. Darwin collected and preserved 1,529 species in jars, and prepared another 3,907 dried specimens of fossils, skins, bones, birds, and corals. Most of these he shipped back to England throughout his voyage, and the vastness of his collections became well known before he returned. His notebooks included over 1,700 pages of descriptions, corresponding in great detail to the collections that awaited him. As a result of training before he left, his experiences onboard the *Beagle*, his correspondence during the trip, and the collections in his possession, Darwin was considered one of the leading naturalists of his day, and he was just twenty-seven years old when he returned to England. In addition, his account of the trip topped almost all other exploration books in popularity. *The Voyage of the Beagle* (1839) was a bestseller in London and beyond.

Darwin's consuming interest in the years that followed his voyages revolved around explaining the great diversity of species of plants, animals, and fossils that he had seen on his voyage. He had filled his notebooks with as many questions as descriptions, and he began to hope that some solution to this problem of species might emerge. He believed, for example, that the many different forms of tortoises on the Galapagos Islands might all be descendents of a smaller number of ancestral forms. Over time, on different islands with slightly different conditions, different species might arise. The same appeared true of mockingbirds on the Galapagos. Additional examples appeared in South America, when Darwin considered the similarities of extinct rodents and armadillos to living species. He eventually imagined a solution to the species problem by considering the possibility that diversity existed because over vast periods of time, species had changed. Other naturalists, including Darwin's own grandfather Erasmus, had suggested as much, but the young Darwin knew of no mechanism that would adequately explain how change in species occurred. He determined that, like Newton, a scientist should seek an explanation in material terms for such a mechanism.

By the mid-1850s, only a few of his closest friends knew that Darwin was nearly ready to present a complete account of evolutionary change by a mechanism he had begun to call *natural selection*. In 1858, Alfred Russell Wallace, another naturalist who had traveled around the world and examined diverse species, wrote an essay describing a mechanism for evolutionary change that closely resembled Darwin's own. Wallace sent the essay to Darwin, asking for his opinion of it. Darwin immediately realized that it was time to go public with his ideas. If he hesitated, Wallace could publish the essay, and Darwin would receive little or no recognition for his life's work. Darwin graciously and strategically helped Wallace by arranging a joint presentation of their ideas to the Linnaean Society in London. They could then publish the mechanism independently, with a shared sense of priority.

From that point forward, Darwin worked quickly to produce a single, fully documented argument for natural selection. In *On the Origin of Species*, first appearing in 1859, he provided details and analogies (mostly to dog and pigeon breeding) that would explain natural selection. As a result of the completeness of Darwin's case, and evidence that he had developed the idea several years earlier, Darwin earned greater recognition at the time and ever since. He had assembled more data to support the mechanism of natural selection and had argued for the veracity of the idea more convincingly than had Wallace. Wallace became properly recognized as a codiscoverer of the mechanism of natural selection, although he made a smaller contribution to articulating the idea.

Darwin explained the mechanism of change in a remarkably simple way. He demonstrated that populations of each species would be capable of growing much larger than any environment could support. As a result, many individuals would die before maturing or having the opportunity to reproduce. Within this struggle for existence, those that did survive and reproduce would be selected by natural conditions as those best suited to the current environment. Those traits that suited individuals to their environment would be passed on to the next generation through biological inheritance—a subject Darwin recognized needed considerably more research. The struggle for existence would be repeated in every generation, so that a changing environment would favor the individuals in each generation that best suited the conditions. Over time, the population's composition would change from individuals that suited one environment at one time to individuals that suited a somewhat different environment a generation or more later. Unlike previous accounts of evolution, Darwin succeeded in offering a fully materialistic explanation for change. The response among most scientists was enthusiastic, but the implications for the place of humankind in this scheme created an ongoing stir in the waters of human controversy, with waves that still crest today.

Exploring Science in a New Nation

When the United States declared independence from Great Britain, its leaders embarked on a new experiment in government. Based on principles of democracy, this government offered an uncertain haven for scientific work. Men like Benjamin Franklin and Thomas Jefferson possessed significant reputations as naturalists, but they sought only minor support for either a federal scientific academy or a national natural history collection. While certain sectors of the public found nature a fascinating subject, the authors of the Constitution could not justify including science as a significant component of the government's role in society. They would not create an institution for science dependent upon government funding.

Naturalists in America tended to follow the debates in Europe with interest, but generally did not actively engage in them, with a few exceptions. Buffon, near the end of his life, had noted the prevalence of small species brought back to Europe from America. Jefferson responded with evidence showing that his home possessed animals with equal, if not greater, vigor. He created a table of comparisons for Buffon, demonstrating that among four-legged beasts, American representatives were larger in several instances. Examples like the extinct wooly mammoth and the flourishing American Bison extended the range of size to compete with the largest mammals anywhere. Such a showing for natural history in the United States, however, hardly constituted a justification for the government to commit regular funds to these studies, which could generally be accomplished by men like Jefferson in their spare time.

The only generally agreed upon role for the newly established government in promoting the work of natural history and philosophy involved issuing patents for new inventions. This practical role ensured that inventors would be rewarded for their innovations, although much of what naturalists did in this period was not patentable: collecting, naming, and classifying species. The practical interests of the new nation lay in expanding its technical capacities and promoting industry that could compete with the established economies of Europe.

Without the wealth of monarchs to support studies of nature, American natural history offered fewer opportunities for aspiring naturalists than they might have found in European institutions during that time. Individuals could not prosper by specializing in botany, chemistry, or even mineralogy. A few men—as a result of their profession in fields like law or medicine—found the time to build collections as an avocation. Following the model of European naturalists and philosophers, they organized a limited number of exclusive groups for the exchange of ideas, such as the American Philosophical Society in Philadelphia.

Some private collections also grew into well-known institutions, building on the public's interest in nature. These collections included William Bartram's (1739–1823) botanical garden and Charles Willson Peale's (1741–1827) natural history museum. These were located in Philadelphia and, not coincidentally, affiliated with the Philosophical Society. Bartram's botanical work included the publication of a descriptive volume based largely on his travels in Florida, the Carolinas, and Georgia. It became a classic in New World botany and was translated into French. Despite this success, Bartram never earned much international acclaim. Peale's museum grew out of a desire to provide a public exhibition of natural history. He did not consider it a research collection for philosophical inquiry into nature and, as a result, the museum has been remem-

William Bartram (1739–1823)

William Bartram developed an interest in botany with help from his father, John Bartram (1699–1777), who was named the Royal Botanist to the colonies by the British royal family. The elder Bartram participated as a member of the leading intellectual community of Philadelphia in the eighteenth century, an auspicious position in society for a man raised on a farm with almost no formal schooling. William Bartram began life with many advantages his father had lacked, including the family's extensive library. He grew up accustomed to having guests such as Benjamin Franklin in the family home.

William Bartram (1739–1823) made extensive plant collections and earned an outstanding reputation as an early American plant explorer and plant taxonomist. (Library of Congress)

Bartram's first work as a naturalist came during travels with his father, who called William "my little botanist" and brought him along to observe and collect specimens. The boy began sketching species, and other naturalists quickly recognized his artistic talent. His sketches, through John Bartram's connections, made their way to Europe. Linnaeus even based some of his descriptions of American bird species on William Bartram's drawings.

Like many other naturalists, Bartram considered a career in medicine, which would provide a steady income and a recognized role in society. Because of his early exposure to natural history, he decided not to enter medical school. At the age of 15, he entered the Philadelphia Academy, at the time an institution in only its third year. It later became the University of Pennsylvania. His instructors included some of the finest classical scholars in the New World at the time. Bartram spent four years there and went on to become a full-time naturalist after attempting careers as a merchant and as a farmer. He failed at both of the latter, but succeeded admirably as an illustrator, collector, observer, and writer. He lived briefly in Florida and traveled throughout the southeastern states.

Bartram eventually moved back to his father's farm in 1778. His brother had inherited it after their father's death, and William lived there the rest of his life. In 1782, the University of Pennsylvania offered him a professorship but he declined, preferring to remain on the farm, where he helped from time to time but took little interest in actually running the place. He kept his focus on natural history. His best known work carried the descriptive title *Travels Through North and South Carolina, Georgia, East and West Florida, The Cherokee Country, The Extensive Territories of the Muscogulges, or Creek Confederacy, and the Country of the Chactaws* (1791). As one of the few active naturalists of the eighteenth century, Bartram's circle of friends included intellectuals from many other fields. Both of the poets Coleridge and Wordsworth referred explicitly to the work of their friend in the poems "Kubla Khan" and *Ruth*, respectively. Bartram came to be known as the first native-born American to devote his entire life to the study of nature.

bered primarily as a precursor to P. T. Barnum's circus. Although Barnum did eventually buy half of Peale's collection, Peale himself established the museum to promote public understanding of nature. Significant attractions, which he recognized in organizing the museum, included the many beautiful and bizarre forms to be found among the plants, animals, fossils, and minerals of North America. Peale was, in part, a Philadelphia entrepreneur who could draw the public into this rather unique American institution and then provide them with a new awareness of natural history.

The efforts of these few individuals, in the absence of a federal consensus for governmental support of scientific work, laid a certain amount of the groundwork for American science. There clearly existed a demand for scientific institutions, both as a result of practical needs for technical and natural knowledge and as a consequence of the interest in nature that was as common among educated American men and women as it was among Europeans. While the U.S. government may not have embraced this demand in the late eighteenth or early nineteenth century, leaders of the new democracy found themselves promoting inquiry into nature in prominent ways.

Perhaps the most significant natural history work done in America in the early nineteenth century resulted from an increasing desire to identify the geographic limits of this mostly unexplored continent. The expeditions led by Lewis and Clark, Ferdinand V. Hayden, Clarence King, and John Wesley Powell had set out to explore the continent. They each brought back new species of plants and animals, descriptions of geological formations, and systematic records of the native people from the unknown center to the far corners of North America. When President Thomas Jefferson began planning a major expedition in 1802, he asked the Spanish government for permission to send a party of men up the Missouri River, which was then controlled by Spain. Jefferson made assurances that the expedition was for "the advancement of geography," and not an attempt to secure a trade route to the Pacific. At the same time, he asked the U.S. Congress for funds to support the endeavor with assurances that it would enhance commerce opportunities. Congress agreed to provide $2,500, and Jefferson asked Captain Meriwether Lewis to lead the expedition. Lewis had little experience or training as a naturalist, but he was a capable and courageous leader. To better prepare for the expedition, he went to Philadelphia and took a crash course in collecting plants and animals as well as in making celestial observations and gathering information about the native people that the group would encounter. His teachers in this brief period were the young nation's finest philosophers and natural historians.

Meanwhile, in 1802, the unexplored territory passed from Spanish control to the French. Jefferson wanted to ensure the success of his planned expedition, so he asked Napoléon about a land purchase. The French ruler surprised Jefferson with an offer to sell the entire Louisiana Territory for $15 million. The American president actually hesitated at the bargain, wondering whether Congress would approve the deal. When Congress enthusiastically agreed, Jefferson completed the purchase. It took most of 1803 to finalize the details, and by then the U.S. was especially eager to find out what it had bought. The boundaries consisted mostly of general descriptions of unexplored lands, though many people believed a waterway might be found to connect the Missouri River with the Columbia. Captain Robert Gray had developed commercial possibilities near the mouth of the Columbia River on the Pacific Ocean, giving the U.S. some claim to what was known as the Oregon Territory. A connection from New Orleans to the Pacific, a "Northwest Passage," would revolutionize trade with Asia.

Lewis shared leadership responsibilities for the expedition with William Clark. Along with twenty-six soldiers, two interpreters, one slave, and an Indian woman (Sacajewea, who was the wife of one of the interpreters), Lewis and Clark made their way up the Missouri River. They left St. Louis in the spring of 1804, reaching the western part of present-day North Dakota by late fall. After spending the winter there, the party followed the remote stretches of the Missouri into the Rocky Mountains. The Lewis and Clark expedition made the potentially difficult crossing of the Continental Divide, from the Missouri River to the Snake River, with remarkable ease. Once on the Snake, they followed that river to the Columbia and reached the Pacific Ocean on November 7, 1805. They spent the winter near the mouth of the Columbia and returned to St. Louis in less than a year. The party separated briefly in the Rocky Mountains to do additional surveying and collecting on their return trip.

The expedition brought back a rather small collection of plants and animals from the newly explored territory. The animal specimens included mostly familiar mammals: mule deer, antelope, wolf, and hare. Several of the specimens were preserved more to the specifications of military explorers than of naturalists, which matched the general composition of the expedition. The list of items brought back included, for example, a "buffalow robe" and the horns of a mountain ram. They also packed up some living specimens on the last stretch of their journey, such as a "squirrel of the prairies," magpies, and a "hen of the prairie."

In the end, the expedition surpassed almost everyone's expectations, yet they had not achieved either of the major goals of the mission. They did not find a continuous water route to the Pacific—as Jefferson had hoped in his appeal to Congress for funding—nor did they bring back extensive collections of new species—as Jefferson had originally explained their mission in requesting Spain's approval. What Lewis and Clark had discovered, however, was a vast network of rivers connecting areas of seemingly endless forests, prairies, moun-tains, and gorges. Each of these areas represented new and previously unimag-ined opportunities for expansion of American wealth: forests for lumber, land for agriculture, mountains potentially rich in minerals, and rivers for power and transportation. While it would take most of the nineteenth century to realize the full potential of these resources, the spread westward had already begun. The example of how Americans could control and industrialize nature was becoming well established in the East and Midwest. As they moved westward, settlers applied the social, economic, and productive machinery to western towns, rivers, and lands.

Conclusion

Throughout the eighteenth century, natural history underwent constant reform with the discovery of new species from around the world. Naturalists working in Europe struggled to keep pace with these discoveries of explorers. Although the contributions of only a few individuals are discussed here in detail, each was part of a larger network of men and women who traveled, collected, and classi-fied the vast array of new species. These people also wrote accounts of the places they visited, providing a broader view of the natural world than had been available to any previous generation of naturalists. Gesner, Linnaeus, and Buf-fon, while somewhat exceptional for the enormity of their contributions, do rep-resent a range of perspectives for studying the natural world. In a different way, Lyell and Darwin provided important explanations for the unity and diversity of natural forms. Their theoretical and observational work relied on the explo-rations and collections of the naturalists who preceded them.

In North America, vast reaches of wilderness lay unknown to the commu-nities of naturalists in Europe. Detailed examination by settlers, trappers, and explorers revealed an abundance of natural resources and unexpected similari-ties to familiar regions. Human settlement began to expand into these new lands, providing even greater incentive for full investigation of the possibilities of the environment.

By the middle of the eighteenth century, an expanding awareness of how humans could shape the environment began to refocus naturalists' questions. Instead of wondering how God had created the world as we find it, they began to question how humans could utilize the abundance of nature. Natural history, natural philosophy, mechanical philosophy, and the economy of nature increas-ingly blended into new structures of environmental science that could exploit natural resources as much as they sought to understand nature.

Bibliographic Essay

Historians have contributed a wide range of synthetic works on the ways humans have explored and understood nature, from ancient times to the beginning of the modern era. The most comprehensive of these is Clarence Glacken's *Traces on the Rhodian Shore: Nature and Culture in Western Thought from Ancient Times to the End of the Eighteenth Century* (Berkeley: University of California, 1976). Peter Bowler provides a useful overview, with greater detail than is possible here, to this period in his *The Earth Encompassed: A History of the Environmental Sciences* (New York: Norton, 1992). Dealing more specifically with the development of science over the centuries before the scientific revolution, David Lindberg provides an excellent and accessible source, *The Beginnings of Western Science: The European Tradition in Philosophical, Religious, and Institutional Context, 600 B.C. to A.D. 1450* (Chicago: University of Chicago, 1992). On the spread of knowledge around the world precipitated by exploration and settlement, consult Alfred W. Crosby, *Ecological Imperialism: The Biological Expansion of Europe, 900–1900* (New York: Cambridge University, 1986); Richard H. Grove, *Green Imperialism: Colonial Expansion, Tropical Island Edens, and the Origins of Environmentalism, 1600–1860* (New York: Cambridge University, 1995); and Keith Thomas, *Man and the Natural World: Changing Attitudes in England, 1500–1800* (London: Allen Lane, 1983). The most readable general introductions to the development of natural history contain less detail than the sources mentioned above, but two remain essential for any interested reader who wants to follow this history and compare the perspective of an historian of science—Paul L. Farber, *Finding Order in Nature: The Naturalist Tradition from Linnaeus to E. O. Wilson* (Baltimore: Johns Hopkins University, 2000)—to that of an environmental historian—Donald Worster, *Nature's Economy: A History of Ecological Ideas*, 2nd ed. (New York: Cambridge University, 1995). The work of Linnaeus and Buffon are also placed in the broader context of the naturalist tradition by Farber in *Finding Order in Nature*.

Colonial science and the development of science in the newly formed democracy remain difficult subjects to explore because of the complexity of exchange between American naturalists and those in Europe. The best sources remain A. Hunter Dupree, *Science and the Federal Government: A History of Policies and Activities* (Cambridge, Massachusetts: Harvard University, 1957), and John C. Greene, *American Science in the Age of Jefferson* (Ames: University of Iowa, 1984). In addition, popular interest in nature in that period constituted an important arena for development of knowledge for a broad audience, as explained by Toby A. Appel in "Science, Popular Culture and Profit: Peale's

Philadelphia Museum," *Journal of the Society for the Bibliography of Natural History* 9:4 (1980): 619–634.

The fascinating debates over how the Earth was shaped are best told by Martin J. S. Rudwick, in *The Great Devonian Controversy: The Shaping of Scientific Knowledge among Gentlemanly Specialists* (Chicago: University of Chicago, 1985). Biographies of Darwin abound, but a recent and readable account of his idea development is given by Adrian Desmond and James Moore, *Darwin: The Life of a Tormented Evolutionist* (New York: Warner, 1991). Janet Browne is writing a multi-volume biography, the first of which covers his childhood and years aboard the *Beagle, Charles Darwin: Voyaging* (New York: Knopf, 1995). For more on botany in the United States, as well as the reception of Darwinism in this country, see A. Hunter Dupree, *Asa Gray: American Botanist, Friend of Darwin* (Cambridge, Massachusetts: Harvard University, 1959). Michael Ruse's account of Darwin's work—and the ideologies that inspired and derived from it—is found in *The Darwinian Revolution: Science Red in Tooth and Claw*, 2nd ed. (Chicago: University of Chicago, 1999).

2

Realizing Nature's Potential

The relationship between humans and nature began to undergo a dramatic change beginning in the early nineteenth century in the U.S. and across western Europe with the emergence of industrialization. The Industrial Revolution created new demands for scientists, some of whom saw the natural world as raw material for industrial processes. Others recognized the potential for a philosophy of nature to unlock the secrets of nature's mysteries. In the midst of that potential, natural historians and philosophers began to reflect on the significance of changes they observed around them, both in nature and in human society. Some of the changes led to cycles of expansion, where demand for resources led to new means of harvesting and transporting goods, leading to even further demand. These cycles began to accelerate in the Industrial Revolution, bringing the relationship between science and the environment into sharp focus.

The adoption of the steam engine as a primary source of power generally marks the beginning of the Industrial Revolution. Various types of steam-driven machines preceded the engine invented by James Watt (1736–1819) in about 1775. In fact, one of those predecessors, the Newcomen engine invented by Thomas Newcomen, had been in use in many parts of the world for decades when Watt made the innovations that distinguished his steam engine as the new standard. Rather than marking a single date as the beginning of the Industrial Revolution, the more interesting features of this important period involved the role of science in industrial advances and the effect of those advances on society in general, and the environment in particular.

More than any specific application, the steam engine revolutionized the processes of production in industries like textile making, agriculture, and transportation. As time passed, changes in each of these industries had even more profound effects on human society and the environment. Humans realized their ability to reshape landscapes and natural communities on new levels. The ability to do so also introduced the possibility that they could instead choose restraint. Ultimately, that choice raised new options in how one might view nature.

As the Industrial Revolution progressed, it took on different forms in the newly independent United States and other parts of the world, especially Europe.

James Watt (1736–1819) was an engineer and inventor in Great Britain during the 18th century. He was best known for his invention of the first practical steam engine, which played a major role in the Industrial Revolution during the 19th century. (Library of Congress)

European societies, for all of their diversity and complexity in the eighteenth and nineteenth centuries, had undergone little widespread change in their processes of manufacturing and agriculture. With the sudden rise of industry, systems of political and economic power that had been in existence for centuries disintegrated, causing hardships for people who depended on those structures. In America, industrialization brought less dramatic changes. The settlement of Europeans

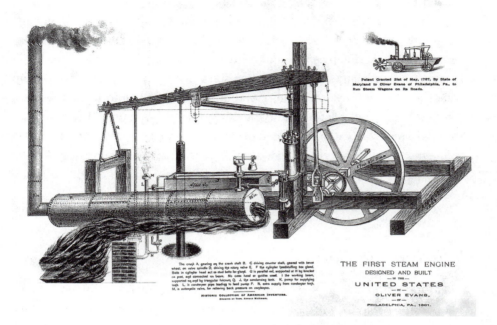

The first steam engine designed and built in the United States. Steam engines provided a new source of energy in the 18th century that quickly became widespread in Europe and the United States. (Library of Congress)

across the Atlantic Ocean brought European cultures in contact with seemingly boundless resources. As a result, American processes of manufacturing and agriculture underwent almost constant expansion and innovation beginning from the first experimental establishment of communities in this "New World." Native Americans had a great deal to teach new settlers about how unfamiliar resources might be used, but as native communities were pushed to the margins—and in many cases destroyed entirely—American settlers developed a new sense of possibilities for industrializing their society. The new opportunities coincided with an expanding awareness of available resources. New classes of workers, manufacturers, and farmers could emerge without the complete realignment of society.

At the same time, the view of nature from Charles Darwin's perspective—as the product of evolutionary change—suggested that humans occupied no special place at all. Darwin and many of his followers suggested that a tree could represent evolution, with many branches and twigs coming from the trunk, yet with humans at the top. This approach retained the connections of the hierarchy, but eliminated the sense that nature could be understood as the linear progression of a single creation. Among the many branches, singling out humans seemed as much like hubris as assuming all creation led to the pinnacle of humanity. Many people reacted negatively to this suggestion, fearing that a loss of uniqueness among all creatures might give way to a widespread loss of adherence to moral and ethical principles. People had long based these principles on the belief in a special relationship between humans and a divine creator. This relationship seemed jeopardized by evolution, creating new and uncomfortable questions regarding the human relationship with nature. Could humans still assume that nature existed solely for their convenience and pleasure? Could they also assume that resources were provided specially for human consumption? Unlike the system described by Linnaeus and his predecessors, the view of nature by the end of the nineteenth century left these questions cloudy, at best.

Clearing the Land

The earliest settlers from Europe began cutting trees for firewood and also to make room for food crops. They quickly cleared those areas closest to their towns and gradually expanded their operations. By the first decades of the eighteenth century, local governments placed certain restrictions on forest cutting to ensure that some trees remained for specialized use. One such use was for ships' masts, which required tall, straight, white pines. In other areas, landowners burned much of the lumber cut from a forest where it lay to enrich the soil.

A sizeable portion of the wood went straight into the stoves and fireplaces of individual homes. In the absence of any other fuel source during the cold New England winters, families heated their homes exclusively with wood. A typical home might burn the wood from a full acre of forest every year. They used a relatively small amount of the wood for construction unless it could be easily hauled to a mill, and while their construction needs could be fulfilled in a year or two, landowners' demand for fuel was continuous.

Settlers processed forest and agricultural products near their homes, and people met their own needs with small milling operations. As demand for more finished products increased, milling operations became more significant. Companies built larger mills near sources of power, such as waterfalls and river rapids, requiring effective transportation of raw goods to that point and from there to the final marketplace. As steam power became more widely available, those lines of transportation became more efficient. Processing could take place near the source of the goods or, more often, near the cities that served as the main market for the finished products and where a larger workforce was available to operate the mills. The appearance of larger mills ultimately made it possible to process more lumber, which made it profitable to cut larger sections of forest every year. The appearance of technology that improved the efficiency of creating a finished product expanded the interest in producing raw materials. Markets for the finished product could be found in urban areas, not only in North America but also in England, across Europe, and beyond.

When settlers cleared forestlands, especially in New England, the remaining soils often proved inadequate for intensive crop production. Land that supported dense forests did not respond well to the conversion to farming. The vast and deep root systems of trees enabled them to grow on uneven, rocky, and relatively steep ground with poor soil. When landowners removed those trees and roots, they often found thin soil that quickly washed down rocky hillsides. Farming on recently cleared land typically entailed several years of labor in order to remove rocks and establish crops that would hold the remaining soil, and that soil required enrichment through the addition of minerals and organic material.

As settlers spread west, they found richer farmland that required far less preparation. They could grow crops more quickly on the level land where glaciers had deposited deep topsoil. Transporting corn and other crops from these areas proved far more profitable. The spread of farming to Ohio and other Midwestern lands took place rapidly in the nineteenth century. Although they had little awareness of the factors that depleted soils of nutrients, farmers recognized the symptoms of depletion in reduced yields. When yields dropped and farms became unprofitable, many farm families moved farther west. In their place, new immigrants sometimes managed to restore soils by adding nutrients from organic material such as manure. Rather than corn, tobacco, and cotton, crops like alfalfa and soybeans could return additional nutrients to the soil, however these were less profitable on the open market.

Transporting agricultural and forest products to cities on the Atlantic Coast became more important both as a result of population growth in those cities and because of decreasing yields from forests and farmlands from Maine to the Carolinas.

Agriculture

The southern states suffered economically after the Revolutionary War because the market for the region's primary cash crop, tobacco, was disrupted by postwar trading changes. Without British support, American shippers faced new difficulties in reaching their usual markets. Great Britain now viewed American products as competitors to products from its remaining colonies, prompting restrictions on trade and leaving American ships to fend for themselves against pirates on the high seas. Tobacco farmers felt the change most acutely. As tensions between the southern and northern states increased in the middle of the nineteenth century, southern farmers suffered greatly when the North established a blockade on trade of tobacco and other southern crops. Southerners had to smuggle cotton out of their ports, some 1,250,000 bales during the Civil War. The cost of this form of trade, which included buying provisions, clothing, hospital stores, manufactured goods, and small arms, depleted the southern economy, contributing to the exploitation of land resources.

Besides the economic problems, southern farmers began to recognize that tobacco cultivation depleted their soil. Although it remained the most profitable crop for farmers, planting and harvesting tobacco year in and year out had unfortunate effects on the land. Alternative cash crops would not grow well enough to make a substantial profit and few markets existed for food crops in a country where much of the population still lived on subsistence farms of their own.

While some farmers succeeded in growing wheat or rice, others turned to cotton. Before the 1790s, only one type of cotton could be grown profitably. This type, known as long staple or sea-island cotton, grew best on the warm, humid, coastal lands of the Carolinas and Georgia. Another type, short-staple cotton, could be grown more widely in the South, but the process of picking the seeds out of the shorter fibers was labor intensive. One person could spend an entire day picking the seeds from a single pound, while someone working a simple roller device that pressed the seeds out of long-staple cotton could process twenty-five pounds in day. As a consequence, even though it was easier to cultivate, families grew short-staple cotton only for limited home use.

In 1793, cotton processing changed forever. A mechanic who had recently graduated from Yale College visited a southern plantation and created a device that could pick the seeds out of short-staple cotton. The mechanic, Eli Whitney, invented a machine that could process up to fifty pounds per day of short-staple cotton. Whitney's cotton gin (the word "gin" being a familiar shorthand form of "engine") revolutionized southern agriculture. The textile industry felt the effects of that revolution far beyond the southern plantations.

An abundant supply of cotton from the South in the early nineteenth century meant an expanding textile industry. Manual operations scattered around the country operated at low levels of efficiency. More efficient and more profitable textile mills required large numbers of workers and a source of power to run the looms. Locating textile factories in larger cities brought them closer to an eager workforce and improved economic returns, especially in New England. Lumber operations had experienced the same improvements from an increased centralization of milling.

As cities grew, the market for food crops also increased, making larger scale agriculture a necessity. By the 1820s, subsistence farming had given way

Replica of the cotton gin built in 1793 by Eli Whitney. (Corbis)

to market farming in New England and the mid-Atlantic States. The rapidly growing populations in these areas and relatively long use of the land soon led to soil exhaustion. Farmers typically solved this problem by moving west. Larger farms appeared around small villages that grew quickly into towns across New York, Pennsylvania, and eventually into Ohio. These transformations in human society multiplied after the introduction of new means of producing food and clothing.

Wherever markets for food, clothing, and other products expanded, people looked for ways of increasing the efficiency of producing these goods. In the nineteenth century, production was linked closely to human labor, but new devices could enhance the efficiency of that labor tenfold or more. A farmer might still control almost every aspect of cultivation, from planting to harvesting, but new tools made it possible to work more land and thus increase the total crop brought to market at the end of the season. Much more than the addition of workers, the innovations in agriculture in the 1820s and 1830s expanded production for the growing eastern markets.

A proliferation of new plow designs in the early nineteenth century gave farmers a range of choices for breaking up the soil. Whether the land had been cultivated previously, consisted of recently cut and cleared forest, or was covered with rich prairie sod, by attaching the appropriate plow to a team of oxen or horses, just a single person could prepare soils for planting (although the work was far from easy). Farmers adopted heavier wooden plows covered with wrought iron, requiring teams of up to sixteen oxen to pull them. John Deere improved the wrought iron plow by welding sharper, smoother steel to the cutting edge, which could slice thick prairie sod more effectively. Instead of scattering seeds on the surface of freshly turned soil, devices capable of depositing seeds in machine-drilled holes helped farmers reduce waste and improve yields. Finally, harvesting crops, which often required that individual plants be kept in alignment for easier processing, became systematically simpler with the introduction and refinement of threshing machines.

All of these improvements in efficiency came in response to a farmer's need to work quickly at certain times of the year. Throughout the history of agriculture, a single family could only work a relatively small plot of land. They prepared the soil in the spring as soon as the weather warmed enough to thaw the frozen ground. Planting had to take place early enough to maximize the growing season, but only after the threat of further freezing was safely past. As a result, there was a limited window for success in planting. Harvesting also had to take place fairly quickly. If ripened crops were not processed in time, they might spoil, rendering a full year's work worthless. In order to increase the scale of agriculture, each of these tasks had to be completed within the same time limits. The ability to plant more seed would be meaningless if a corresponding increase in harvest could not be accomplished. In the midst of rapid industrialization, each aspect of a farmer's task received corresponding attention from equipment innovators. Larger scale farming also meant large-scale production. The most successful farm implement company of the nineteenth century, created by Cyrus McCormick in 1841, became an early model of American manufacturing. Machines and equipment became a vital component of farm economies and processes.

These changes in farming practices enabled farm families to spread across North America. In each place they settled, these families transformed acres of forest or prairie into cropland. Plots of land, from 40 to 160 acres, began to fill up with corn, wheat, cotton, and other crops. Survival required hard work for every member of these families, and they could not afford to waste any part of the slim returns for their efforts. Some years, rainfall proved inadequate for the crops they had planted, and other years their fields flooded and entire crops were destroyed. Each season brought a new lesson, and each lesson taught farmers to make better use of their land. This meant maximizing the amount of land in cultivation, cutting remaining groves of trees and plowing under remnants of prairies. The creation of farmland, from the East Coast to the Midwest to the Great Plains and beyond, resulted in a massive transformation of the environment that naturalists began to examine only late in the nineteenth century.

Transportation

If coordinating the processes of planting, growing, and harvesting marked the first obstacle overcome to serve the expanding American population, the next obstacle involved transportation. Moving goods from increasingly far-flung farming communities to the bustling cities could take days or even weeks. The losses resulting from slow transit included time away from other essential labor on the farm for working men and draft animals alike. Roads were rough and deeply rutted during certain seasons of the year, making them almost impassible. Improvements in wagons started a revolution in transportation. With larger wheels and sloped storage beds, these wagons overcame most ruts and prevented spillage. Still, the basic construction of roads changed little before the middle part of the nineteenth century, due in part to the enormous expenses involved in building roadways with easy grades and fine, compacted material that would remain level in changing conditions. Some states planned and built roads along major transport routes, and the federal government commissioned a "National Road." In general, however, governments did not take responsibility for developing a network of roads that would facilitate transport for private enterprise. A few exceptional cases demonstrate the staggering expense and slow progress of road building. The federal government completed the National Road, which ran from Baltimore to Wheeling, south of Pittsburgh, in 1818 at a cost of approximately $13,000 per mile. Costs increased as it was extended farther west. The state of Pennsylvania spent $500,000 to build the Lancaster Turnpike in the early 1790s. Rather than fund transportation routes with tax money, most users of early roadways paid for the privilege in the form of tolls.

Water transport had obvious advantages where rivers could carry goods downstream. Since the larger cities lay near the coast and generally along riverways, boats and rafts were put to regular use. Because streams and rivers were apt to change dramatically with the seasons, however, this mode of transportation required skilled navigators in many cases. Shifting sandbars and other unseen obstructions created hazards to water transport. In addition, moving goods upriver to the growing towns in the interior of the country was almost impossible. In 1807, Robert Fulton succeeded in building a steam-driven paddleboat, which traveled 150 miles upstream from New York City to Albany in just 32 hours. Similar voyages made on placid sections of rivers had sparked enthusiasm since about 1790, although the widespread use of these expensive boats seemed unlikely on the majority of the nation's riverways.

Fulton and his partners managed to navigate several rivers by steamboat, including the Ohio and the Mississippi from Pittsburgh to New Orleans. Nicholas Roosevelt, one of Theodore's granduncles, led the first voyage on this stretch, leaving Pittsburgh in late September of 1811, reaching New Orleans on January 12, 1812. This voyage established regular steamboat runs by Fulton's company between those cities. In 1819, steamboats began operation on the Great Lakes. Because these trips constituted interstate commerce, the U.S. Supreme Court eventually ruled that Fulton and his partners would have to submit to federal regulation, thus ending their monopoly. This legal setback was minor compared with the reality of the safety record of steamboats. As many as one-third of these vessels were lost in major accidents before 1850, mostly on western rivers. Fulton's competitors were not immune to the safety hazards, but they did manage to develop lighter steamboat designs, which could navigate shallower waters.

The innovations in steam-driven water transport, such as shallower designs, made it feasible to consider opening new waterways to navigation. By digging out or dredging the shallowest rivers, steamboats could reach new destinations. Even more entrepreneurial was the proposal that canals might be dug across the American interior to link larger and previously unconnected waterways. Inspiration for the idea came partly from Europe, where canals in some areas had served commerce for centuries. However, digging a new system of canals in the U.S. would be a massive undertaking. It began with short links, mostly within cities like New Orleans, Charleston, and Boston.

Fewer than 100 total miles of canals existed in 1817, the year work began on the Erie Canal, which would link the Great Lakes to New York City. The canal stretched 363 miles from Lake Ontario to the Hudson River, via the marshes of the Mohawk Valley and the Finger Lakes of central New York State. The designers and builders of the Erie Canal overcame problems of construction by developing new solutions. They planned a waterway forty feet wide at the top and twenty feet wide at the bottom, and four feet deep. To keep water at a navigable depth throughout, a system of locks could be filled and drained to move boats in both directions. They completed the canal in 1825, along with a section running north from Albany to Lake Champlain. The total cost ran over $10 million, but the profits from traffic on the canal quickly exceeded the cost, given the savings of canal transport compared to overland travel. The success of the Erie Canal inspired construction in other states, but nowhere else did canals prove as economical. In Pennsylvania, for example, mountainous terrain meant a need for many more locks, which added expense and reduced the efficiency of transport by that route. In most areas, canals were soon eclipsed by improved road construction and the introduction of railroads.

The growth and success of railroads became enmeshed in the social and geographical changes taking place in America. As agriculture moved west, railroads made it possible to transport agricultural products east. Conversely, manufactured products and finished goods from the East could reach expanding agricultural towns economically, making life in those places less rural than anyone a century before might have dared to imagine. The market for manufactured products thus increased, further spurring the growth of cities where manufacturing was concentrated. As cities grew, so did the demand for agricultural products. The cycle of growth accelerated throughout much of the nineteenth century in America, especially in the North. Railroads played a vital role in that growth, although much historical evidence suggests that even without railroads, agriculture would still have expanded westward and industry would still have increased. The reasons for growth in those areas included population increases and the opening of new markets, and wagons or canals could have filled the role of railroads into the 1890s. Nevertheless, railroads did play a crucial role and created certain markets of their own.

Railroads heightened the demand for improvements in steam engines, which ran their locomotives. At world expositions in the 1850s and 1860s, American locomotive designs won awards for the advances in technology. With eight wheels, they were lighter and more flexible than their European counterparts, equipped to handle the steeper grades and sharper corners of American roadbeds. They also received equal parts of praise and criticism for the elaborate and—some said—garish decorative metal and woodwork that adorned their locomotives. The designs were seen as distinctly American and contrasted with the European locomotives, which ran on rail lines in countries with different social and economic priorities. The railroads and locomotives in the U.S. presented a unique statement about American ingenuity and style, which many believed still existed only in the shadow of European dominance.

American railroads enjoyed additional advantages beyond the eager markets at both the eastern and expanding western ends of the line. The iron industry underwent a series of innovations in production and corporate structure that improved the quality and availability of iron for tracks and trains alike. Steel production became a seamless process that began in iron mines, continued through smelters, and ended in factories where other machines fabricated finished goods. The tracks consisted of metal rails attached to wooden ties, and abundant forests across much of the eastern part of the country provided the thousands of log ties needed for mile after mile of track. Transportation also transformed the lumber industry, allowing for more centralized milling practices.

Railroads established as early as the 1830s consisted mostly of short lines, connecting major ocean and river ports with one another or with manufacturing districts in major cities. Along the East Coast in particular, railroads connected Boston with Providence, Lowell, and Worcester. In New Orleans, a railroad connected Lake Pontchartrain directly with the Gulf of Mexico, allowing boats to avoid the long trip through the Mississippi River delta, a shallow and complex waterway in the best of times.

When railroads reached the Middle West in the 1850s, the level terrain made it possible to connect more cities more rapidly than ever before. Building a railroad from Baltimore to Ohio meant crossing the mountains by following rivers and smaller valleys. Once in Ohio, however, railroads could connect Cleveland and Cincinnati with Detroit, Chicago, and St. Louis. From each of these cities, railroads spread in every direction, reaching ever more remote markets with eastern goods and bringing back the agricultural and forest products of Wisconsin, Minnesota, Iowa, Indiana, Illinois, and Missouri.

The Utility of Natural History

While Charles Darwin represents a somewhat exceptional case of a naturalist developing mechanistic theories, many others derived similar inspiration from earlier natural historians and from their own travels without reaching the same conclusions. In most instances, naturalists worked toward more practical goals of understanding how plant and animal distributions could be manipulated for human benefit. In America, many states established natural history surveys that would examine the conditions of forests, rivers, streams, and ponds to determine how fish and wildlife populations remained in balance. The resources of lumber and soil also constituted areas of significant interest and concern. George Emerson's report of the trees of Massachusetts, which had inspired Henry David Thoreau, was one example. Another, written by Vermont attorney and statesman George Perkins Marsh, had an even greater impact on concern for the natural world.

In 1846, George Emerson published a report on the trees and shrubs of Massachusetts. The state had commissioned the report in 1837 as part of a zoological and botanical survey that would promote agriculture. Emerson found time to work on the survey during breaks from his work as an educator and president of the Boston Society of Natural History. The report became a catalog of the woody plants of the state, for which he relied heavily on Asa Gray's descriptions and the publication of another botanist, William Oakes. Emerson introduced the catalog with a preface that listed the problems associated with deforestation. He acknowledged that the natural growth and decay of forests enhanced soils, and that hillsides from which trees had been cut would quickly be stripped of their soils by erosion. He also suggested that hilltops left barren of trees would lack the means of conducting electricity from clouds to the ground that might stimulate the fall of rain. This hypothesis that lightning was essential to enabling rain to fall does not conform to modern ideas of meteorology but, for Emerson, it pointed to the double jeopardy of stripping hills and valleys of their forest cover. Trees provided protection from both heat and cold in summer and winter.

In this official report, Emerson limited his discussion of the natural beauty of the forests he described, but he did admit that a complete removal of aesthetics was impossible for him. Indeed, his sketches of leaves and trees reflected the simplicity and beauty of the plants he depicted. He also attempted to limit the amount of technical language in his discussions. His main point, then, was to provide useful explanations of the amount of forested land available in the state and the value of those forests for a variety of purposes. For example, most of Emerson's discussion focused on the most "noble" and valuable timber species, such as pine, elm, and oak. He paid less attention to poplars and other fast growing species of limited economic importance.

Emerson calculated the total value of wood products for construction in a given year to be just over $2 million. If the value of wood that went into the building of ships could be added to that, the sum would be even greater. Forests also provided products such as nuts from the nut-bearing trees, and sugar from maple trees. The use of wood for fuel amounted to an additional value of between $45 and $50 per household. Finally, railroads also used an enormous amount of wood for construction of rail lines. All of these uses, combined with a recognition that forest users did very little planting to replace the trees they cut, added up to an impending shortage. The eventual shortage of lumber would, in Emerson's estimation, result in a serious loss of manufacturing in Massachusetts.

A study of timber cutting in Vermont led George Perkins Marsh to make a more sweeping statement about the impact of humans on their environment. Marsh's expertise on the topic of the environment came from his experience as a political administrator and a careful observer. Since he grew up in Vermont, he was—like Thoreau in Massachusetts—familiar with its natural areas and the changes caused by human activities. Even more than Thoreau, Marsh developed an active concern that those changes might ultimately prove destructive not only to nature, but also to human society. His interest in understanding the changes in nature had more to do with his concern for humanity than a romantic or moral concern for nature. Trained in literature and the law, he was less qualified than many naturalists of his day to undertake this project, but also because of his training, Marsh was uniquely qualified to take a synthetic approach to describing the problems as he saw them and then offering solutions. Even before the German zoologist Ernst Haeckel coined the term ecology in 1866, Marsh emphasized the same need for study of living beings and nonliving factors together in their environment. He added a suggestion that humans played a significant role in

George Perkins Marsh (1801–1882)

George Perkins Marsh wrote the classic account of the relationship that developed between humans and nature in the nineteenth century, *Man and Nature: Or, Physical Geography as Modified by Human Action* (1864). Its later significance has suggested to many readers that Marsh was a prophet of conservation. In fact, he was neither a scientist nor a wilderness lover. He never visited the great western forests or places like Yosemite Valley. Instead, Marsh made observations that reflected his deeper concerns about economics and changing uses of resources in New England. Along the way, he pointed out many themes that would resonate with conservationists for decades to follow. The book inspired people to consider the source of natural resources and the way human societies had altered their surroundings.

George Marsh's 1864 work Man and Nature: or, Physical Geography as Modified by Human Action *provided a descriptive basis for concerns that grew into the conservation movement in the United States. (Library of Congress)*

Marsh was born in 1801 in Woodstock, Vermont, near the Connecticut River. His father participated actively in politics and served as a member of Congress during the War of 1812. He followed in his father's footsteps, first studying law and later being elected into Congress before becoming the U.S. Minister to Turkey in the late 1840s and 1850s. He also served as Abraham Lincoln's ambassador to Italy beginning in the 1860s, and remained there until the end of his life. These government posts suited Marsh. Closer to home, he served in the Vermont State House and used his architectural skill to design the state capitol building.

nature and titled his book on the subject *Man and Nature: Or, Physical Geography as Modified by Human Action* (1864). He did not neglect the expertise of naturalists along the way. He corresponded with Asa Gray in planning the book, and he called for new research to be done by qualified naturalists. Marsh was most optimistic that scientific inquiry could soon answer questions that, at that time, were unsolvable using historical evidence.

In observing nature during his lifetime, Marsh concluded that humans caused the overwhelming majority of changes in plant and animal relationships. Many of these modifications promised to be quite harmful in the long run. He focused on four broad areas. He wrote long chapters on plant and animal species, the woods, the waters, and the sands. The introduction and conclusion laid out the fundamental argument of his book, and examples throughout the book testified to Marsh's expansive knowledge of the natural world, from his native Vermont to the far corners of the planet, including the Arctic, Europe,

He also contributed to the design of the Washington Monument in the District of Columbia. As the Vermont Railroad Commissioner, he investigated and reported on the irresponsibility of corporations. As an art collector, he amassed one of the finest collections of prints and engravings in the country at the time. He spoke twenty languages and traveled around the world. His excellent memory enabled him to recall virtually everything he had ever read, seen, or heard. He developed this ability early in life, when an eye illness made it difficult for him to read regularly himself, and forced him to retain information that was read to him or offered by other means.

In the late 1850s, as the state fish commissioner, Marsh undertook a study of the declining fisheries in Vermont and elsewhere. This report sparked him to write more on the changes taking place in the natural world. As a boy growing up in Vermont, he had enjoyed a firsthand look at the rich variety of forested areas. He appreciated the balance of nature and adopted William Paley's perspective of natural theology. Comparing his later observations in the Mediterranean region and the European Alps, Marsh saw that humans accelerated natural forces like erosion. He did not want to see America repeat the errors of the Old World. As a statesman and a designer, he recognized that humans had been architects of misfortune, and he hoped that trend could be reversed by foresight and with technical skill.

Marsh's *Man and Nature* caught the public's attention almost immediately, and it had an influence on forestry. He was critical of the purely economic focus of timber cutting and called for broader concern about the losses resulting from unrestrained harvests. He described forestry as the most significant disrupter of nature's balance. Charles Lyell read the book and became convinced that humans have a greater effect on nature than he had generally acknowledged in his own work. It was several decades, however, before the conservation movement developed around similar concerns but within a different social context. Marsh's work was not fully acknowledged as part of this movement until the 1930s, fifty years after his death.

South America, and parts of Asia and Africa. Of the changes caused in all of these places by human activity, Marsh wrote that, in far too many cases, members of his own species consumed and wasted the natural bounty that surrounded them. In the vastness of the United States, this consumption and wastefulness seemed far from causing any real damage. He feared, however, that humans could eventually reproduce the same pattern on this continent that had left large areas of other continents completely devoid of forests and other living things.

At its most basic level, Marsh believed in the stability and balance of nature. Stability, however, could not be expected to persist in the face of consumption and waste. Once upset by human activity, he did not believe that a safe future could be assumed. Referencing the evolutionary awareness that naturalists at the time proposed, he stated quite clearly that the earth had not adapted to the ways of humanity. While species and environments could change over generations in ways that favored the long-term survival of well-adapted species, human modification of nature happened too quickly and was unprecedented in the planet's history. As such, Marsh concluded, "The ravages committed by [humans] subvert the relations and destroy the balance which nature had established between her organized [*sic*] and her inorganic creations . . ." (*Man and Nature*, 42). He went on to suggest that nature could and did retaliate against humanity. For example, when people cut forests and left land barren, nature responded with "deluges of rain" that washed away the dry soil. Similarly, drought could turn deforested land into deserts in a generation.

Recognizing the importance of vegetation, Marsh demonstrated his extensive knowledge of the diverse roles of plants in nature. Agriculture and forestry, while essential to human society, immediately created change in an area due to the fact that harvests removed virtually all of the "products" of plant growth from the land. In nature, plants would die where they had grown and "rot upon the ground" (*Man and Nature*, 54). Generally speaking, plants had profound effects on their local environments, changing the chemical composition of the air, penetrating into the ground with their roots "to greater depths than is commonly supposed," and holding the soil together with root filaments. Plants also affected the humidity of an area, by taking water out of the air and using that water in their own tissues. In recounting this importance, Marsh commented as much on the damage done when plantlife was inhibited, counting the advantages of vegetative growth. He noted that changes in vegetation caused by humans included the transfer of many species from one continent to another. From America, he listed maize and the potato as being extremely valuable to European economies. His opinion of another American crop, tobacco, was quite different. Feeling in part culpable for the spread of this plant around the world, Marsh wrote, "I wish I could believe, with some, that America is not alone responsible for the introduction of the filthy weed, tobacco, the use of which is the most vulgar and pernicious habit engrafted by the semi-barbarism of modern civilization upon the less multifarious sensualism of ancient life . . ." (*Man and Nature*, 58). He might have been an ally of twentieth-century general surgeons.

With respect to animal life, Marsh exhibited less expertise and nowhere near the same sense of conviction, but he referred to a range of species that faced changing fortunes with the modifications of human activity. In those middle years of the nineteenth century, the American bison represented the only native animal that could rival humans in transforming their environment. Marsh's opinion of these animals changed before publishing the next edition of his book just ten years later when he saw that humans had virtually extinguished bison by the 1870s. As for other large herbivores, the deer, elk, and mountain sheep, he acknowledged that their numbers had never been great, and the carnivorous animals that hunted them were even fewer. This made the domesticated animals brought by European settlers even more significant. Cattle and sheep quickly dominated the American landscape. Marsh summarized, "[I]t is evident that the wild quadrupeds of North America, even when most numerous, were few compared with their domestic successors, that they required a much less supply of vegetable food, and consequently were far less important as geographical elements than the many millions of hoofed and horned cattle now fed by civilized [humans] on the same continent" (*Man and Nature*, 74).

The spread of human culture across America or any other continent necessitated the clearing of forests. Marsh highlighted that many civilizations cut trees not for fuel nor for building materials, but simply to open land for other uses. Trees, he acknowledged, provided little food and tended to preclude the growth of edible plants generally. In the process of clearing land, one civilization after another succeeded in changing the local climate of their new homes. Marsh described examples of climate change in European countries in previous centuries. In Italy, forests cut from higher elevations resulted in increased winds through the valleys below. In subsequent years, houses had to be built with heavier roofs; where once straw sufficed, tile and stone roofs became the norm, due to the winds.

When people cut forests, more serious changes in temperature and humidity also resulted. Marsh repeatedly noted that naturalists and meteorologists understood weather patterns only very poorly. In many cases, they observed that trees offered more surface area than bare ground for the collection of radiant heat from the sun. This, they reasoned, could lead to higher temperatures over a given area. On the other hand, less of that heat reached the soil, so lower temperatures near the ground could also be observed in forested areas. For the same reason, naturalists saw that forests retained higher humidity than deforested areas, so soils stayed moist. Levels of humidity, however, could not be related directly to rainfall, another variable in the complex weather picture. Some evidence suggested that forested areas received greater amounts of rain, with trees acting like wicks to pull moisture from the air. When forests were cut, however, even slight amounts of rain heightened the risk of erosion. Marsh assembled a variety of landslide accounts that attested to the consequences of deforestation, especially in hilly or mountainous regions.

With so much speculation as to the actual effects of human modification of physical geography, Marsh might be seen as a mere worrywart. He even wondered whether the amount of carbon released as gas by the destruction of forests might somehow alter the atmosphere. The conclusions he reached, however, demonstrate the significance of his work in this period. He regularly referred to the most reputable sources of natural knowledge, including relatively recent books by trained naturalists like Gray, Lyell, and Darwin. He also consistently balanced his own remarks with broader comparisons to the comments of other observers from around the world. Marsh took special pains to provide quantitative evidence wherever possible. For example, with the question of carbon gas in the atmosphere, he could compare differences between carbon from forests and that from other sources of decaying vegetable matter. Although he could not calculate exact amounts released, he did conclude that the effect of trees would be "infinitesimal" (*Man and Nature*, 125). Marsh could not predict the significance of carbon dioxide as a controversial product of industrialization.

The variety of approaches Marsh adopted gave him a perspective that transcended natural history and tied scientific knowledge to a wide range of practical concerns. Realizing the complexity of the issues he discussed, he argued that people should be cautious about adopting hard and fast rules for addressing environmental concerns. In most cases, Marsh wrote that "the evidence is conflicting in tendency, and sometimes equivocal in interpretation," so most naturalists were inclined to rely as much on historical and comparative information as on facts that were more directly obtained (*Man and Nature*, 158). Even where they could collect certain data, contradicting data might eventually come to light, requiring further support either for the initial conclusion or the subsequent contradiction. Examples from European countries, where humans had groomed and modified nature for generations, offered guidance to Marsh. Referring to those countries, he asserted, "Let us be wise in time, and profit by the errors of our older brethren!" (*Man and Nature*, 198).

Looking Back at Progress

In the midst of tumultuous changes during the first half of the nineteenth century, we see that Americans and Europeans alike adopted an exploitative view of nature. The vast resources of North America, for example, became an illustration of the abundance available for human use. Those people whose ancestors had come from Europe a generation or two earlier embraced the potential for expansion, and they were joined by more immigrants every year. Whether one chose to grow crops, cut trees, raise livestock, prospect for gold, or run a shop, the possibilities for wealth seemed nearly limitless. In addition to natural resources, new means of production based on steam engines and other mechanical devices made some forms of labor less arduous. Expanded transportation by road, rail, and water meant that goods could reach ever-larger markets, allowing profits to continue to increase.

Each of these advances carried a price, or at least one writer adopted that view in the middle of the nineteenth century. Henry David Thoreau lived almost

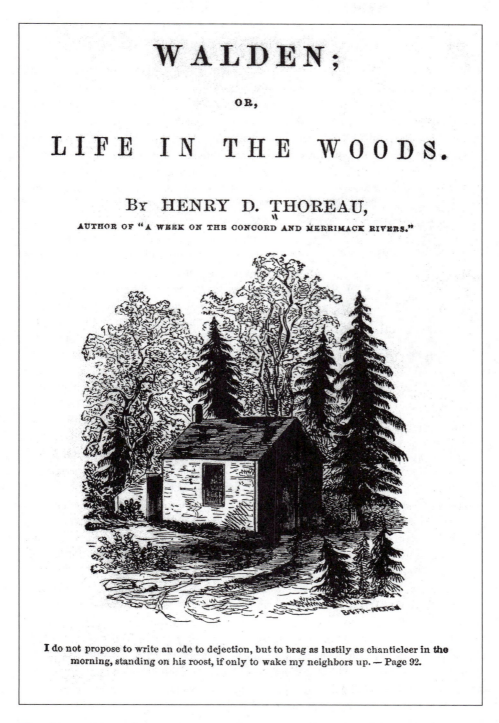

WALDEN;

OR,

LIFE IN THE WOODS.

By HENRY D. THOREAU,

AUTHOR OF "A WEEK ON THE CONCORD AND MERRIMACK RIVERS."

I do not propose to write an ode to dejection, but to brag as lustily as chanticleer in the morning, standing on his roost, if only to wake my neighbors up. — Page 92.

The title page from the first edition of Henry David Thoreau's Walden: or, Life in the Woods. *Thoreau wrote of his experiences and thoughts during a two-year period when he lived in a tiny one-room cabin he had built by the shore of Walden Pond near Concord, Massachusetts. (Library of Congress)*

Henry David Thoreau (1817–1862)

In 1845, Henry David Thoreau built a shack on Walden Pond, costing him twenty-eight dollars and twelve-and-a-half cents. He lived there for a year, making observations of plant and animal life and watching the changes of the seasons. He called himself an "inspector of snow storms and rain storms"(*Walden*, 15). Some days he would walk twenty or thirty miles in the woods and along the roads near Concord, Massachusetts. Although he was only twenty miles from Boston, one of the busiest cities in America in the nineteenth century, Thoreau found himself in a remarkably rural setting near Walden Pond.

Portrait of Henry David Thoreau (1817–1862), 19th-century author and transcendentalist. (Library of Congress)

Thoreau's life during that year modeled simplicity in a society that he saw as increasingly and unnecessarily complex. His food consisted of Indian meal, rice, rye, beans, and whatever his friends might prepare for his frequent visits to their homes around dinnertime. The simplicity he enjoyed was not a complete "back to nature" experience in which he subsisted entirely on the provisions of the pond, the forest, and nearby streams. In fact, Thoreau made almost no attempt to limit himself to subsisting on nature. Instead, he hoped to do without the complexities of life as he saw it being played out in the towns and cities around him so he could reflect more intently on his own basic needs. Those needs, ultimately, could be met as much by human society as by natural providence, but his approach was unique at the time for the effort he put into considering which needs would best be met by humanity and which could be provided from nature. He summed it up famously: "I went to the woods because I wished to live deliberately, to front only the essential facts of life, and see if I could not learn what it had to teach . . ." (*Walden*, 85).

Thoreau was born in Concord in 1817 and lived most of his life there. He was given the name David Henry, but reversed the names and was always known as Henry. His father was a storekeeper, trader, and pencilmaker, while his mother raised the four children: an older sister and brother, Henry, and a younger sister. Henry attended Harvard College beginning in 1833, and returned to Concord in 1837 as a schoolteacher. The next year, he opened his own school, but he gradually spent more of his time writing and traveling around the area on foot. Later in life, he traveled to Maine, Cape Cod, Canada, and Minnesota.

his entire life in Massachusetts, around Concord not far from the bustling seaport of Boston. He was a schoolteacher, a writer, a naturalist, and a philosopher. This combination made Thoreau a keen observer both of nature and of human society. He possessed something of an entrepreneurial spirit, but not in the sense of industrialists who looked for ways to make money by opening new markets for trade or finding new sources of wealth in nature.

Thoreau admired the work of naturalists who came before him. Linnaeus, in particular, provided inspiration for Thoreau. The Concord naturalist read Linnaeus's "Economy of Nature" and copied long passages into a notebook. He educated himself by this reading but also recognized the challenges to understanding the world around him through the eyes of men who lived a century or more before him on another continent. As a consequence, Thoreau set out to do just what previous generations of naturalists had done. He went out into nature to make his own observations.

For Thoreau, going out to find nature meant something quite different than what we might expect. The wilderness of the Rocky Mountains, for example, lay mostly beyond his reach. He did not actually think of remote wilderness as representative of nature. Instead, he was most interested in the workings of nature just outside his own door. Rather than collect new or exotic species from faraway places, this naturalist in Concord wanted to know the place where he lived, and to know it intimately. As a kind of experiment, he decided to build a small house—a shack really—not far from the homes of his parents and friends. He was not trying to "get away from it all." He actually visited and dined with friends and family on a regular basis. He was even thought of as a freeloader, inviting himself over for dinner when his own stocks ran a little short, or when he went walking all day and found himself hungry and "in the neighborhood" at mealtime. These details about Thoreau's way of life illustrate his unique approach to natural history.

What Thoreau discovered by approaching nature in this way was that human society contrasted with the economy of nature that Linnaeus described and attempted to explain. Many people had begun to assume that since nature's economy could readily be made to accommodate human needs, the best way to proceed was to make accommodations wherever and in whatever ways possible. This consequence of Linnaeus's work did not match Thoreau's philosophical views. Thoreau began to suspect that Linnaeus had matched his theology to his study of the natural world in ways that served humans better than they reflected the workings of nature. Consequently, while he admired the way his predecessors conceived of nature as thrifty and efficient, Thoreau recognized that disruptions and change over time call for more complex explanations. From his own experience in New England, he saw that forests consisted of many of the same species that the earliest European settlers had encountered, but gone were the stands of white pine. Individual trees had formerly reached 250 feet tall; maples 15 feet in circumference were once dwarfed by oaks big enough that it took five or six adults touching fingertips to reach around the giant trees. Those forests had teamed with life, including bears, pumas, wolves, turkey, white-tailed deer, and squirrels by the thousands. Such forests had disappeared by Thoreau's day. Extensive forest lands returned to New England after soils were depleted and farmers moved west, leaving lands to be reclaimed by native species, and many new ones too. Thoreau saw these changes firsthand and also read of them in Emerson's report on trees and shrubs in Massachusetts. Thoreau's views were shaped in part by the society he shared with New England merchants, farmers, and industrialists, but also by the work of naturalists like Lyell and Darwin. He challenged Linnaeus's views of nature on his own terms and believed they would need replacement in changing times.

Conclusion

Industrialization transformed nature in North America, and agriculture, transportation, and an increasing awareness of the continent's resources each played a role in that transformation. Natural historians and philosophers participated in this transformation by contributing to the technological advances, but they also played a role in assessing the changes to landscapes and wildlife species. In ways that appear quite different from the work of ecologists and environmental scientists today, naturalists of the early nineteenth century continued collecting and describing the elements of the world around them. Without the support of wealthy patrons, academic centers, or government institutions, they resembled casual amateurs more than serious professionals in their work, yet the information they gathered began to demonstrate the ways in which the environment around human societies was affected by farming, forestry, and industry. Although they might not have foreseen it, the material they compiled also provided a foundation for further comparison by subsequent generations of naturalists.

Bibliographic Essay

Recent scholarly histories of the expansion of human culture across North America and the resulting exploitation of natural resources are few and far between. A number of more local case studies represent the phenomenon, but do not consistently capture the general trend toward settlement. Broad discussions of industrialization and the place of nature appear in Peter Bowler, *The Earth Encompassed: A History of the Environmental Sciences* (New York: Norton, 1992); Joseph Petulla, *American Environmental History* (San Francisco: Boyd and Fraser, 1977); and Donald Worster, *Nature's Economy: A History of Ecological Ideas* 2nd ed. (New York: Cambridge University, 1995). Worster also discusses Thoreau at length. For more on technology in this period, see John F. Kasson, *Civilizing the Machine: Technology and Republican Values in America, 1776–1900* (New York: Penguin, 1976). George Perkins Marsh's classic is *Man and Nature: Or, Physical Geography as Modified by Human Action* (Cambridge, Massachusetts: Harvard University, 1864).

William Cronon provides an excellent review of changes in New England in the colonial period in *Changes in the Land: Indians, Colonists, and the Ecology of New England* (New York: Hill and Wang, 1983). On the opposite coast of North America, a case study of exploitation of resources in California from a scientist's perspective is Raymond F. Dasmann, *The Destruction of California* (New York: Macmillan, 1965).

Several histories of forestry and American efforts to manage timber and other resources provide insight into the role of politics and expansion initiatives. These include: David A. Clary, *Timber and the Forest Service* (Lawrence: University of Kansas, 1986); and Harold K. Steen, *The U.S. Forest Service: A History* (Seattle: University of Washington, 1976). A longer cultural historical view of America forests is given by Michael Williams in *Americans and Their Forests: A Historical Geography* (New York: Cambridge University, 1989). An important account of water management in the western United States is given by Donald Worster in *Rivers of Empire: Water, Aridity, and the Growth of the American West* (New York: Pantheon, 1985).

3

From Natural History to Government-Sponsored Science

T he contributions of nineteenth-century naturalists such as Charles Darwin marked a period of transition in natural history. The activities of description and classification became distinct from efforts to explain phenomena by natural processes. Emphasis now shifted to viewing nature as an arena for the "survival of the fittest." The new approach matched the rise in the social prominence of science during the second half of the nineteenth century. Many biologists turned to the laboratory sciences of physiology and morphology, including a new trend for marine laboratories where specimens from the sea could be studied in abundance. Yet the laboratory and museum work that reflected the rising professionalism of science offered only one dimension of the answers to questions about the relationships of plants and animals to the environment. The traditional field-work of natural history continued to excite researchers beyond the lab. In an industrializing world, questions about the relationships of species to their environment became increasingly pertinent and practical.

Naturalists interested in studying the broader environment of plants and animals often had a very practical aim. With industrialization on the rise, naturalists hoped to gain a better understanding of the ways in which nature's bounty could be not only harvested, but also harvested profitably for generations to come. Given the warning of observers such as George Perkins Marsh, they recognized the prudence of making resources sustainable rather than simply depleting them and moving elsewhere to do the same. For these practical naturalists, nature represented a limited fund of resources for which all species could compete. By the late nineteenth century, any naturalist could recognize that the conditions of plants and animals at any given time were in large part the product of historical processes. By the same token, changes initiated in the present could affect humans and the environment well into the future. These realizations were the products of a larger transition in natural history.

In the broader social arena, industrialization and urban growth made nature seem insignificant to many people. They moved to the cities and found work in the factories. For the first time in American life, agriculture began to take a backseat to industry. As they became increasingly detached from nature, most Americans retained a view that nature was a vast storehouse of resources. Human action would be too small to have much of an impact. However, the permanence of nature became a topic of debate as industrialization began to overpower certain resources and to transform particular places. The realization that nature could change resonated with a Darwinian view of evolution and even struggle. The question for some observers became "how much" rather than "whether" humans could modify nature.

Naturalists took one aspect of Darwin's contributions—his economic analogies—to new levels of sophistication. Debate over the extent to which humans could transform nature had special implications for the consumption and depletion of natural resources. For example, more studies became focused on declines of fisheries and dwindling of forests. Increasing attention to the workings of economics inspired scientists to take economic principles more seriously in numerous ways. They considered checks and balances between species, and they thought more quantitatively about exchanges of energy within populations.

The use of economic analogies in natural history represented only one point of contact between science and society during this period. Studies of change in nature found a parallel in human society in the late nineteenth century when the U.S. entered a reform era of *progressivism*. This period of reform included efforts to clean up politics that had fallen into the hands of special interest groups. Big business faced new levels of regulation in order to protect the public, creating penalties for unfair practices. Reformers hoped to place ideals of efficiency, expertise, and order at the heart of political processes. These ideals matched scientific standards, which raised science— and especially scientific management—to new levels of prominence. Science appeared to open "unlimited opportunities for human achievement" (Hays, *Conservation*, 2).

Optimism for the future relationship between science and society helped bring scientific research into the federal government at an unprecedented pace. Since the way society viewed nature and human action in this period intersected with the rise of economic metaphors and social reform, some observers believed that other parallels in understanding might also be exploited. Nature might serve as a guide to improvements, rather than just a coincidental reflection of certain aspects of human society. The search for these connections became an important starting point in progressive reform, and it had implications for the way nature would be viewed and protected into the early twentieth century.

Nature Examined in Fine Detail

Inspired by Darwin, the American naturalist Stephen Forbes took a view of nature that included many analogies to economics. Forbes wrote a paper in 1887, "The Lake as a Microcosm," that became famous among later ecologists because of its careful discussion of the connections among the species in a single lake. He termed the lake a "microcosm" to emphasize the broader relationships between various species throughout nature as a whole. As a scientist, he wanted to do more than observe and record the species he found. He hoped to provide some explanation of their interactions. He described the relations of species within a "community of interest." While later generations of ecologists expanded the meaning of the word "community" to designate a carefully defined set of species in a given area, Forbes used this term simply to denote the "common" or shared interest of species he could observe.

Forbes recalled Linnaeus's sense of an order in nature, although he did so without reference to a divine plan. Forbes embraced a Darwinian view, where the beneficent power of natural selection adjusted the multiplication and destruction of species to promote a common interest. Nature directed the survival of any given species, leading unconsciously to a kind of optimum state in the common interest of all the species connected to the particular species. He saw balance as an attainable goal of nature, but being a result of natural occupation rather than divine design. Forbes made use of an analogy to human systems of monetary economy to demonstrate the process of nature's order. He wrote, "Just as the thrifty business man who lives within his income will finally dispossess his shiftless competitor who can never pay his debts, the well-adjusted aquatic animal will in time crowd out its poorly-adjusted competitors for food and for the various goods of life" ("Lake as a Microcosm," 549–550). Nature could thus attain its own balance.

Forbes studied the role of different species in the common interest of nature. He asked, for example, how birds feeding on insects might create patterns of increase and decrease in insect populations. He also wondered if birds would change their preferences based on the abundance of different insects. Thus, his work had practical implications that, as a government natural history researcher, came with the territory. He could not afford to study nature in all its complexity without regularly boiling it down to facts that might be useful to farmers or commercial fishermen. Forbes wrote an article entitled "On Some Interactions of Organisms" in 1880, in which he described the way practical considerations would affect how naturalists conceived of the classification of species. He wanted to collect information that would enable him to express the "*functional* relations" of species—those activities that modify the numbers,

Stephen Forbes (1844–1930)

In the late nineteenth century, Stephen Forbes easily ranked as the most recognizable ecologist in the United States. Forbes was born in 1844, and around the time he entered college, the country entered into the Civil War. He did receive one year of formal training at Beloit Academy in southern Wisconsin, but he then left to serve in the military. During the war, he took a special interest in medicine, and when he returned to civilian life, Forbes headed to Chicago to study at Rush Medical School. The realities of medical school differed markedly from his experience in the field, and the financial cost proved more than his meager savings from the military could provide. He dropped out before finishing even his first course.

With almost no formal training but a growing interest in natural history, Forbes began collecting specimens and decided on a career in teaching. In 1871, he entered the Illinois State Normal University. (The word "normal" here refers to the training of teachers.) His skills as a naturalist earned him enough recognition to be selected as the new director of the Illinois Museum of the State Natural History Society just one year later. John Wesley Powell, another Civil War veteran, vacated the post when his own reputation enabled him to move on to Washington, D.C.

Despite having attended three different formal institutions, Forbes could still be described as mostly self-taught in the field of natural history. Yet by 1875, he was named a professor of zoology at the Normal University he had attended, and in 1884, he became chair of the Department of Zoology at the University of Illinois. The teaching and research involved in these university positions did not occupy all of his time, for he served simultaneously as director of the State Laboratory of Natural History beginning in 1877.

Also in 1877, Forbes moved from publishing primarily on topics of teaching (no doubt a requirement of his positions at the museum and the Normal University) to publishing on the interactions of plants and animals. He had grown particularly interested in the food habits of birds and fish, especially when quite different species made use of the same food sources. Forbes took special interest in the concept of the balance of nature, and his use of the concept inspired others to refine and investigate its meaning.

When he died in 1930, Forbes was still best remembered for an article he wrote early in his career. The article, "The Lake as a Microcosm," introduced naturalists to thinking of habitats as relatively closed systems. The word "microcosm" suggested to many later ecologists that larger and larger groups of species could be understood by examining the workings of more limited systems. The usefulness of such a suggestion became a source of much debate, in which Forbes himself never took much interest.

habits, and distribution of other species, especially those of potential economic value. He described how life histories—the product of those functional relations—showed how certain species were more intimately connected by mutual interactions than those species that might be more closely related to one another. This suggested a synthesis of Linnaean and Darwinian views of classification along with ecological features of natural selection.

Embracing a thoroughly Darwinian view, Forbes described the "laws of life, the reality of the struggle for existence, the appearance of variations, and the frequent inheritance of such as conduce to the good of the individual and the species . . ." ("Interactions of Organisms," 5). Forbes also recognized that a species would not maintain its numbers if it could not find sufficient food. He noted that an insect that depended on a specific plant species for its food would starve if the plant was not available. Insects influenced their own food supply, creating a balance between insect numbers and available food, and the balance could swing back and forth. He added that this argument would hold not only for insects or even birds, but for all kinds of animals. In these cases, the ideal adjustment matched the reproductive rate of each species to its food supply. The species that served as food would normally produce a surplus sufficient for the species it supports. For him, attaining this balance was not a teleological or designed ideal but rather a product of Darwinian struggle. He went on to illustrate, using general examples, the possible oscillations that a given species might experience, and might simultaneously produce in other species.

Forbes concluded from these general examples that the inorganic features of the environment served to establish the ultimate limits of species increase. As a practical result of these conclusions, Forbes indicated that the most injurious insects were likely to be those that could subsist on a diet of mixed foods and that had few enemies among the bird species of an area.

According to Forbes, humans determined in many cases where to leave the "primitive natural order of life" undisturbed and where that order could be improved by substituting artificial arrangements. Agriculture served as the source of purported improvements in nature. For example, humans do not raise apple trees for the sake of raising more apple trees, but rather for the product of the trees; they would happily cut away whatever part was not necessary to the harvest of apples. Without passing judgment, Forbes acknowledged this as the human way of interfering in the whole economy of every plant and animal. Natural economy, left alone, would replace each individual with another individual of the same kind. Forbes observed that "the principal general problem of economic biology is that of the discovery of the laws of oscillation in plants and animals, and of the methods of Nature for its prevention and control" ("Interactions of Organisms," 16). To succeed in that control, Forbes called for a thorough and intelligent natural history survey. Anything less would be inadequate, and possibly even worse. Only facts about nature would suffice. Many naturalists embraced this view during the late nineteenth century, especially those employed by the U.S. government to place practical and economic biology on a foundation of fundamental knowledge.

Economic Studies of Nature: Wildlife

By the end of the nineteenth century, the U.S. government had grown significantly and the number of agencies devoted to science also increased. Large tracts of public land in the West meant a need for administrative services and, under the influence of progressivism, an administration meant a need for science. A relatively small group of individuals earned positions of prominence and ultimately contributed even more to the study of nature. C. Hart Merriam, who became recognized as a naturalist with broad vision, ascended to the head of the government's primary natural history survey. Spencer Baird began as a naturalist specializing in birds and reptiles, and ultimately became the first head of the U.S. Fisheries Commission. John Wesley Powell, the Civil War veteran and leader of multiple western expeditions, pushed for the creation of the Bureau of Ethnology and became its director. He also simultaneously took the job of director of the Geological Survey. Gifford Pinchot helped transform a small administrative forestry bureau into the U.S. Forest Service, which became a model for government administration of public resources.

The early work of government biologists focused on the study of mammals, birds, and plants to provide expert advice to farmers and policy makers. There was a real tension in that work, however, because naturalists like C. Hart Merriam and his brother-in-law, Vernon Bailey, wanted to focus on identifying and collecting new species while simultaneously recognizing the more practical aspects of their work. They wanted to study the economy of nature without expectations from the government that their work would benefit the nation economically.

Merriam examined the geographical distribution of species, an example of studying the economy of nature. This work led to the concept of "life zones," an idea that gained wide respect from scientists and conservationists alike in the early twentieth century. Merriam developed his scheme for classifying habitats as part of his intensive study of the San Francisco Mountains near Flagstaff, Arizona. He found a consistent relationship between elevation, temperature, and precipitation. Merriam observed that as one moved to higher elevations, temperatures tended to drop and precipitation increased. Thus, from the desert areas at elevations below the town of Flagstaff, one could climb into the mountains and pass through climates that progressed from hot desert to dry forest to cool forest to cold arctic. His work built upon the ideas of explorers in the previous century who had noted climatic changes as one traveled from the equator to the North and South Poles.

Merriam also demanded action to protect particular species. He and many other ornithologists feared that many bird species would eventually be lost to extinction due to overhunting for commercial markets. Hunters killed birds by

Zoologist and naturalist Clinton Hart Merriam was the first head of the Bureau of Biological Survey, the forerunner of the U.S. Fish and Wildlife Service. (Library of Congress)

the thousands to sell their meat and plumage. In addition, farmers customarily killed a wide range of bird species that they believed attacked crops and cut into the narrow profits of farming. Ornithologists, however, insisted that protecting birds had a potential benefit connected to their natural history. Birds could limit insects and other farm pests to an even greater extent than their own consumption of the crops. Understanding the balance of these potential benefits became known as the study of "economic ornithology." Such an understanding required a closer examination of multiple species, as Forbes had begun to demonstrate.

The Department of Agriculture created a Division of Economic Ornithology and Mammalogy, partly in response to scientists' desire for the funding of basic research, believing that scientists would address problems in natural history that affected farmers. The government hoped that the agency could help farmers in identifying the birds and pests that proved most harmful to crops and would find ways of exterminating them. However, Merriam had little interest in serving farmers. Instead, he hoped to maintain a comprehensive natural history focus and immediately worked to have the word "economic" stricken from the agency's name. Although he was deeply interested in the economy of nature—in the sense that Forbes and other ecologists had come to think of it—he resisted the pressure placed on him by the Department of Agriculture to conduct studies that would benefit farmers economically. He wanted his agency to be free to conduct natural history studies according to their own priorities.

Merriam succeeded briefly in distancing his agency from the primacy of practical questions, which became known as the Division (and still later the Bureau) of Biological Survey. Merriam's investigations included the creation of floral and faunal distribution charts for much of the country. By giving the bureau a limited budget, the Department of Agriculture maintained fairly low expectations for practical knowledge. By the turn of the century, however, the demand for practical results from government-funded science forced the Biological Survey to once again focus on problems that had a direct bearing on the nation's agriculture. In the early twentieth century, the federal government called upon the Biological Survey to investigate crop damage caused by insects and rodents as well as livestock damage caused by carnivorous animals.

Economic Studies of Nature: Fisheries

Spencer Fullerton Baird's career took him from the study of birds, lizards, and snakes to the administration of the nation's fish resources as the director of the U.S. Fisheries Commission. Although he imagined that he might continue natural history studies in his government post, Baird faced constant political pressure and tension over how the government would regulate fisheries—an industry with great economic importance in the late nineteenth century. He took as his model George Perkins Marsh's fishery survey of Vermont. Baird agreed that each state should be responsible for regulating its own fisheries. The federal government would restrict its role to studies of food habits of various species and the

Spencer F. Baird was a naturalist and zoologist who served as secretary of the Smith-sonian Institution as well as the first commissioner of the U.S. Fish Commission in 1871 (now the U.S. Fish and Wildlife Service). (Corbis)

promotion of improved management of waterways for the simultaneous benefit of the industry and the fish. As one of his earliest contributions as head of the federal fisheries, in 1872, Baird predicted the collapse of fishing in Rhode Island and Massachusetts due to each state's permissive netting regulations. When the fisheries did not collapse, however, Baird found his credibility in jeopardy and his role as an advisor weakened.

The U.S. Fisheries Commission had to answer to Congress for continued funding, and, although in 1872 Congress granted $8,500 for scientific investigations, Baird realized that his agency would have to provide greater service to the country or else see its budget cut entirely. Congress demanded practical results from the use of taxpayer money, so the commission began to investigate and promote fish culture, the stocking of rivers, streams, and lakes with fish eggs or newly hatched fish from federally operated hatcheries. The opportunity to study the growth and development of a wide range of fish species combined with the economic promise of increasing the number of fish in the country's waterways appealed to both Congress and Baird's agency alike. By 1887, Congress had increased the commission's budget to $185,000. In the years after 1890, the agency devoted less than ten percent of the total budget to scientific research, meaning fish culture had become the federal government's primary objective.

Economic Studies of Nature: Geological Survey

John Wesley Powell served in the Civil War, advancing to the rank of Major and tragically losing his right arm in the battle of Shiloh. After the war, he put his considerable knowledge of natural history to use as a professor at Illinois Wesleyan University and as director of the Illinois Museum of the State Natural History Society. Within a few years, Powell gave in to his desire to explore the western U.S. and began a series of expeditions that ultimately secured him a career as one of the federal government's leading scientists.

In 1868, Powell began to seek financial support from Congress with the endorsement of Joseph Henry as director of the Smithsonian Institution. He began with extensive studies of the geography and geology of the Southwest, leading the first expedition down the Colorado River through the Grand Canyon and, along the way, becoming intensely interested in the Native Americans living in the side canyons and plateaus of the region. In 1879, therefore, Powell proposed the creation of a federal Bureau of Ethnology, which he would direct, in order to collect artifacts and record the lives of these people. Powell hoped that the bureau would be part of a larger scientific agency funded by the government.

Powell also conducted explorations that might inform irrigation practices in the West with scientific knowledge. Others had acknowledged that rainfall was limited on the prairies west of the Mississippi River, but many people continued to hope that agriculture would eventually thrive on those broad, level plains with their deep topsoil. These people believed that replacing native prairie

Major John Wesley Powell with Tau-gu, Great Chief of the Southern Paiute. Colorado River Valley, ca. 1873. (Corbis)

grasses with a regular cycle of plowing and planting would stimulate more rainfall. Powell's survey addressed the practicality of farming on the Great Plains and beyond, attempting to gather more basic natural historical information about the region as well. By the time he completed his *Report on the Lands of the Arid Region of the United States* in 1878, Powell had become convinced that the land use laws developed in the eastern states were inadequate for the West. Following the example of the communities of Mormon settlers in Utah and the Native Americans in Arizona and New Mexico, he concluded that large tracts of land would be required for ranching and raising crops. In addition, the government would have to manage and protect water, the most precious resource in the West, from individual exploitation.

Powell's conclusions served as an example of how science could dictate policy, and opponents of governmental science believed people and policies should be served, rather than driven, by scientific knowledge. As a result, many officials in Washington continued to oppose government involvement in the direct funding of scientific research. They remained skeptical about the practical value of spending taxpayer money on basic research. Government could more effectively direct funds toward practical questions. The belief among many officials that basic research should not be publicly funded prevented the creation of a federal scientific agency. Powell eventually had to settle for a combined Geological Survey and Bureau of Ethnology. He brought the two together when he became director of the former and retained control of the latter in 1881. His success as a scientific administrator derived from his ability to please Congress by continually focusing on solutions to practical problems, but refraining from taking a short view of such problems. Rather, Powell insisted on broader, long-term approaches that ultimately included more comprehensive scientific agendas than Congress might have envisioned. By constantly weaving basic and practical questions, he achieved enormous gains in natural history research without invoking the criticism of taxpayers or their representatives in Washington. It was a fine line that not every scientific administrator could walk.

Economic Studies of Nature: Forestry

Gifford Pinchot began to oversee forestry in the U.S. in 1898 as head of the Division of Forestry in the Department of Agriculture. Among agency directors, he had the most extensive scientific training of his day. The son of a wealthy timber family, Pinchot went to Yale University, then on to the National Forestry College in France. There he learned silviculture, growing forests through scientific management to be harvested as crops. From his father's experiences of clear-cutting forests, which left behind eroded hillsides, Pinchot learned that American forests could not be sustained without proper management, and he believed the government should take responsibility for large tracts of land to ensure their long-term profitability. He began to describe the government's mission as using forests for the greatest good for th number of people over the long run.

When Pinchot took up his post in 1898, he had 123 employees. years, that number had grown to over 1,500, and the Division of Forest the U.S. Forest Service, created by Theodore Roosevelt in 1905. Pinchot conceived the Forest Service in order to improve the federal government's management of its increasing land holdings. In 1891, President Benjamin Harrison set aside 13 million acres as "forest reserves," which stopped the private exploitation of those lands. Six years later, President Grover Cleveland designated another 21 million acres for forest reserves. This land fell under the jurisdiction of the General Land Office, which had no expertise in forest management. Pinchot complained that the lands and his own knowledge of silviculture were wasted by this administrative arrangement. Roosevelt took action, naming Pinchot as chief of the new Forest Service with administrative power over the nation's 159 forest reserves. The president also designated national forests totaling over 100 million acres during his tenure. Pinchot would manage these as well.

Because Pinchot had the unflagging support of Roosevelt and because Congress saw him as the administrator of an agency that could potentially make money for the government, he never had to beg for funds as did C. Hart Merriam and others at the time. The Forest Service could depend upon the wealth held in its lands, selling permits to harvest timber and graze cattle on rangelands at its own discretion. As early as 1872, for the first time Congress appropriated $10,000 to protect timber growing on public land from private cutting and sale. In the late 1870s, the annual budget for protecting public timber went from $2,000 to $2,500. The amount continued to increase until the end of the century when the total reached $290,000. Over a period of seven years beginning in 1898, Congress provided an average of $200,000 per year to the Division of Forestry. Once Pinchot took control of the Forest Service, the annual budget from Congress exceeded $1.5 million.

Even as the government invested in protecting public forests, revenues from timber sales and grazing permits began to escalate. In 1905, the Forest Service received $73,276 from these sources, and the following year the amount increased tenfold to over $750,000. However, this money did not go toward forest research or administration. That money was given to local governments for public schools and road building. The Forest Service needed its own budget for road construction, and in 1916, Congress approved a $10 million program. The growth of these appropriations demonstrated the luxury of administering economically valuable resources with the support of both the executive and the legislative branches of government, which became a legacy of the Forest Service rarely duplicated by other agencies.

Controlling Nature: Wildlife

The U.S. government took different approaches to the management and control of its agricultural, forestry, and fisheries resources. Depending on the features of a given resource and the industry that had grown up around it over the previous century, progressive management involved choosing among various alternatives as to how the resource would be controlled. The government managed the ranching industry, but did not manage livestock or the rangeland itself. Wildlife went largely unmanaged on the open range. At the same time, the government managed fish but not the rivers where they lived or the fishing industry that harvested them. In the forests owned by the U.S. government, the Forest Service sought to manage both the trees and the timber industry. This spectrum of administrative options provided opportunities and challenges to scientists working within the ranks of the government's resource management agencies.

Even more so than administrators like Merriam or Pinchot, Vernon Bailey represented the daily duties of government field naturalists around the turn of the century. He worked for the Biological Survey from 1887 (when he was only twenty-three years old) until 1933. Typical of many naturalists of that era, Bailey never earned an academic degree. In fact, he had little formal education beyond a log schoolhouse in Elk River, Minnesota, northwest of Minneapolis. What he lacked in formal schooling, however, Bailey made up for in diligent fieldwork. He became the Biological Survey's authority in small mammal trapping. His skill in identifying both plant and animal species alike was unquestioned.

Throughout his career, Bailey examined animal habits and habitats, trapped and collected specimens, and worked to promote a wide appreciation of nature. At the same time, he protected agricultural practices—such as sheep and cattle ranching as well as dairy farming—from the depredations of wild carnivores, especially wolves and coyotes. By 1905, the bureau established a program for predatory animal control that catered to western ranchers, while it also gave the Biological Survey a stronger role in what most taxpayers could identify as an important public service. Compared to efforts that supported agriculture, the basic natural history work of the bureau seemed superfluous. The program set up by the Biological Survey became a centralized bounty system, where federal taxes paid hunters and trappers regular salaries in addition to individual bounties. These government employees tracked down livestock killers as well as any other carnivores. Annual reports partially took the place of animal pelts as evidence of the program's success. However, the agency did retain comprehensive scientific studies as well.

The Biological Survey fulfilled the charge set by the ranchers and conducted fundamental scientific investigations. Government naturalists blurred the line between practical results and fundamental data by going into the field with

Trappers and hunters in the Four Peaks country on Brown's Basin, Arizona Territory. Two Crab Tree boys, their father, and the dogs and burros they hunt with. January 1908. (Corbis)

a broad agenda of examining and describing plant and animal species. Their reports ultimately contained a mixture of information that would satisfy a variety of audiences. In 1907, Bailey reported on the habits of wolves, which ranchers considered "vermin" because they killed livestock. His stated mission was to help ranchers and government agents trap and kill the animals more effectively, but Bailey's practical explication of how to control wolves included additional information on their habits and their effects on other species.

In 1908, Bailey described the government campaigns to kill more vermin—wolves and other livestock enemies—and for the first time, he used the term "predatory animal." He enumerated the wolves and coyotes killed in or adjacent to national forests. The actual total might have been twice the official estimate, considering the number of animals in surrounding areas. The federal government's mission was taking shape. It would destroy wolves and other predatory animals throughout the western United States and its territories using the federally coordinated bounty system. During the previous year, government bounty hunters had killed over 1,800 wolves (mostly in Wyoming) and 23,000 coyotes, resulting in an estimated saving of $2 million to the livestock industry. Despite these reports, western ranchers soon grew critical of the bounty method of destroying livestock killers. They realized that, even with federal administration, fraud was undermining effective extermination. Bounty collectors sometimes duped government officials with doctored-up skins and often collected multiple bounties. Livestock owners again called for a federal program to take over the extermination efforts more directly and systematically.

The demand for additional government participation meant an increase in the Biological Survey's involvement, both in administration as well as a source of scientific expertise. Congress granted a small appropriation for "experiments and demonstrations" in the control of animals that endangered livestock. This outlay of funding for scientific study placed the bureau directly in charge of more practical work: the destruction of wolves, coyotes, and other injurious animals. Studies of food habits directly fulfilled the agency's mandate to study the economic relations of animals that affected human activities. The situation became even more urgent with the involvement of the United States in the war in Europe, when food conservation became part of national defense. Concern for livestock continued to be the primary argument for ongoing predatory animal control. The Biological Survey expanded control in the 1910s and 1920s with the use of poison. This work, less directed toward understanding life histories of predatory animals and therefore of significantly less scientific value, became the focus of criticism from scientists outside the bureau.

Predatory animal control had relatively inconsequential effects on game. Populations of wild prey such as deer might benefit from reduction of their natural enemies, but the Biological Survey made no systematic effort to achieve this benefit for game. The agency did, however, provide the most regular source of information on game in the United States. The bureau kept a limited set of game statistics, as well as a listing of game officials and organizations. The latter lists far exceeded any actual data on game, pointing to the expansion of effort in an area where relatively little was known. Most of the officials listed were political appointees of state governments who could contribute little to a systematic study of game conditions. In their administrative positions, they enforced the laws on the books and doled out bounties to varmint trappers. They placed the greatest emphasis on the protection of game birds and songbirds.

Biological Survey naturalists represented the federal government's best effort to study and improve wildlife conditions throughout the country. Ornithologist T. S. Palmer wrote regular progress reports, citing new legislation, court decisions, and expenditures in federal game protection. Much of the work in the first decade of the twentieth century aimed at helping the public understand game laws. Working within the Department of Agriculture, he seemed especially attentive to the needs of farmers by recognizing their role in game protection and the importance of downplaying potential conflicts between domestic and wild animals.

In 1909, C. Hart Merriam described the state of the nation's bird and mammal resources for the National Conservation Commission. He included among the country's assets the "insectivorous and game birds, our splendid game animals, and our native fur bearers." The birds offered value as food, and many also helped to destroy insects and rodents. Some state and local governments protected large game mammals, particularly elk and deer, which faced an uncertain future after a period of unprecedented destruction in the nineteenth century. A small number of game preserves and national parks provided refuge for big game, and according to Merriam, the enactment of wise legislation would ensure their proliferation. The Biological Survey chief even speculated that deer and elk might eventually be domesticated and raised on small tracts of land that were not otherwise fit for cultivation or grazing by cattle and sheep. In the past, hunters killed big game without regard for the maintenance of a breeding stock. "Betterment" of game populations required absolute protection during breeding and migration periods, regulation of the number of animals killed and the means of killing, and a captive breeding program that would ensure a healthy stock. These recommendations, while based on a scientific understanding of the requirements of big game, demanded a broad response at all levels of government, from local to federal. Protection efforts meant increased control over particular species and their habitats.

Controlling Nature: Fisheries

The government also worked to control natural conditions for the benefit of society through the stabilization of the fisheries industry. In 1887, Spencer Baird became the director of the Smithsonian Institution. Marshall McDonald then succeeded him as head of the Fisheries Commission. McDonald differed from Baird in both training and temperament. Trained as a military engineer, he ran the commission with great bureaucratic efficiency, but with little insight into the needs of the administrators and researchers working below him. Under his leadership, the commission significantly expanded fish culture practices. This focus on utilitarian research continued to suit the desires of Congress and the general public.

Understanding fish and their habitats took a much lower priority than producing more fish that could be harvested by commercial fishermen. Although the government had built many dams on the country's rivers and streams for irrigation and flood control, the implications for migrating fish, such as salmon, went unstudied. In annual and multiyear cycles, fish traveled downstream from their breeding grounds to coastal waters, where they spent much of their lives feeding before attempting to return upstream to lay their eggs in cold streams, some-times hundreds of miles inland. Dams blocked fish migration and, while fisher-men and biologists alike realized that fish could not return to their native breed-ing areas, most fish culturists working for the U.S. Fisheries Commission blamed declines in the fishery industry on overfishing and poor regulation by the states. Lacking scientific understanding of the habits and natural history of most migra-tory species, fisheries officials could more easily displace the blame.

Without careful study of fish migration and life histories, states were unable to identify the need to change regulatory practices. The full impact of dams in contributing to the declines went unrecognized for decades. The states, the federal government, and the fisheries industry could not adequately examine the extent and source of problems facing migrating fish and the corresponding fisheries. As a result, the institutions continued their separate policies and blamed the others for these problems.

By 1900, the Fisheries Commission spent over $200,000 each year on fish culture, despite a lack of evidence that the practice of transporting eggs or hatch-lings could increase the numbers of mature fish caught in subsequent years. Because this made the commission an easy target for criticism when the industry showed signs of continued decline, officials defended fish culture even more stri-dently. As one component of this response, the agency demonstrated the process of collecting and hatching fish eggs in tanks and ponds. Fisheries officials set up demonstrations at major international expositions, including the Columbian Exposition in Chicago in 1893 and the Lewis and Clark Centennial in Portland in 1905. Individual fish hatcheries were also equipped with demonstration ponds and placards. Despite these promotions, scientists outside the agency grew criti-cal of its utilitarian goals and the inadequacies of natural history knowledge of migrating fish. In 1910, Stephen Forbes joined other naturalists in calling for com-prehensive surveys of aquatic life. Rather than focus on individual species and the economic outcomes of hatchery procedures, Forbes hoped to see broader inves-tigations. He concluded that releasing young fish into the environment without an understanding of the environment was neither scientific nor practical but rather, "simply ignorant" (quoted in Taylor, *Making Salmon*, 214). The response to this critique, once again, was to defend fish culture more carefully by promoting its direct goals of increasing the number of fish in rivers, streams, and lakes.

Controlling Nature: Forestry

New approaches by the Forest Service involved conservation in the sense that traditional clearcutting practices often destroyed the long-term vitality of entire

A clearcut area of a forest. Clearcutting is a harvesting method in which all the trees in a given area are cut down at one time, leaving behind a field of stumps, broken saplings, and rotten logs. A once-common technique, it is criticized by forestry experts for reducing the biological diversity and resilience of a forest. (U.S. Fish and Wildlife Service)

areas. By controlling how the timber industry harvested trees on public land, the government could make improvements in the use of public resources. Pinchot believed that through a scientific approach, silviculture, forests could produce more wood over longer periods. To better manage the nation's forests, Pinchot established research locations, following the Department of Agriculture's model of agricultural experiment stations. By the early 1910s, stations existed in seven states, including the Forest Products Laboratory in Madison, Wisconsin. These stations also matched the continued enthusiasm for experimental and laboratory science in this period. In Madison, researchers studied the properties of wood and searched for ways to improve the processes that transformed trees into manufactured goods. The goal of this laboratory work pointed directly to economic savings by producing more products from fewer trees.

The experiment stations extended Pinchot's adherence to using science for guidance in planning resource use. The stations promoted techniques that would improve the efficiency of silviculture: fire prevention, replanting trees in rows after harvest, and fighting various diseases and insect infestations. Like other federal experiment stations set up by the Department of Agriculture, the Forest

The Gold King mill processed timber from forests near Telluride, Colorado. Nine-teenth-century mining required lumber for buildings and timbers for shoring up tunnels. To slow the clear-cutting that often resulted from the needs of miners and farmers, Congress passed the Timber Cutting Act on June 3, 1878. (Western History Collection F13267, Denver Public Library)

Service remained oriented toward highly practical goals. Researchers at these stations attempted to model their work after experimental work done in laboratories elsewhere while also functioning in the service of the public forests.

Conservation versus Preservation

John Muir dreamed of going to South America, following in the footsteps of Humboldt as Darwin had. After making his way to California in the late 1860s, he was among the first who would explore the Sierra Mountains and Yosemite Valley in scientific detail. Muir had taken courses at the University of Wisconsin a few years earlier and excelled in botany. He also studied the geological theories of glaciers proposed by Louis Agassiz. In the Sierra Mountains of California, he compared what he had learned in classes to an unstudied landscape, and he proposed theories of his own to explain what he found. Upon seeing striations in

John Muir (1838–1914)

When John Muir founded the Sierra Club in 1892, he brought together California's leading advocates for nature. The club represented the best of Muir's dreams for a simple relationship with nature, and he led members on annual trips into the Sierra Mountains. As a result of those trips, and the extensive writings of Muir, by the turn of the century he had reached a wide audience, explaining his particular dreams about preserving wilderness especially in unique and remote places.

Muir was born in Scotland in 1838 and at age eleven moved with his family to Wisconsin. There, he helped his father build a farm out of wild lands that had never seen crops. He also spent his spare time exploring a new world of trees, swamps, rivers, and hills with his brothers. Muir had never known wild land in Scotland but, through reading, he had come to admire it. Now he found himself surrounded by nature. However, his father kept him very busy and, as a young man, John Muir possessed extraordinary mechanical and inventive skills. He was an asset on the farm and in the factory, and might have been a mechanic for the rest of his life except that an accident in 1868 left him temporarily blind. During this period of blindness, Muir considered the meaning of his life thus far and found that his devotion to machines had restricted his appreciation of nature. He devoted himself again to the study

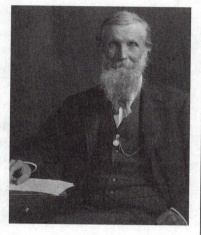

As a naturalist, explorer, and author, John Muir was a major figure in the preservation movement in the United States during the late-19th and early-20th centuries. (Library of Congress)

of his natural surroundings, beginning with a one-thousand-mile walk from Indianapolis to the Gulf of Mexico. From there, he took a ship to California and began his lifelong study of and struggle to protect the forests and mountains of that western state.

Muir wrote many books about his experiences in nature. They included an account of his childhood, his walk to the Gulf of Mexico, and his many hikes in California. In these books, Muir emerged as a philosopher of nature in a sense quite different from natural philosophers like Newton or even Darwin. Muir's philosophy focused particularly on the place of humans in nature, on how nature had been or might be transformed by human action, and whether such transformations ought to be resisted. He lived very simply in those years, in a cabin or outdoors, in and around Yosemite Valley. His writing occupied him, bringing him some income.

The loss of Hetch-Hetchy Valley in 1914 under a reservoir that would serve San Francisco capped a successful career with a potential sense of failure. However, he kept his hopes high to the end. Muir died of pneumonia later that year, with the unfinished manuscript of a book on his travels in Alaska beside him.

Sierra Club Magazine: A cover of Sierra, *the magazine of the Sierra Club. The subtitle, "Exploring, Enjoying and Protecting the Planet," gives a concise summation of the Sierra Club's mission. (Sierra Club)*

granite walls, often the result of ice grinding across a rock surface in a constant direction, and finding erratic boulders scattered on valley floors, Muir assumed he was seeing evidence of glacial action. He published these observations along with a complete explanation of the glacial history of Yosemite Valley and the action of glaciers in the Sierra Mountains. While other naturalists denied the significance of glaciation in Yosemite, including the head of the California State Geological Survey Josiah D. Whitney, Muir's uniformitarian descriptions eventually prevailed.

Based partly on his scientific studies and partly on his personal convictions about the beauty and value of nature, Muir became a leader among those in the United States who hoped to preserve natural places. By the 1880s, Muir and his friends in California hoped to see Yosemite Valley become a national park, protected by the federal government from mining and logging. The Yellowstone area of Wyoming had become the world's first such park in 1872, and Muir believed his home in California deserved equal protection. In 1890, he succeeded when the federal government established Yosemite National Park.

Muir helped to form the Sierra Club two years later. Early members included scientists David Starr Jordan, who was the president of Stanford University, and Joseph LeConte from the University of California. Like Muir, they believed that, with scientific support, preservation could become a national priority. With substantial financial and media backing from Sierra Club members Robert Underwood Johnson, a publisher, and Warren Olney, an attorney, they made a formidable social alliance. Muir soon realized, however, that the Sierra Club's mission was not always perfectly clear.

In many cases, Sierra Club members advocated conservation of resources, which would protect the supply of nature's raw materials from quick exploitation and ensure future generations of their use. *Conservation* became the special mission of many government administrators of public land, such as Gifford Pinchot. Pinchot's belief in management for the "greatest good for the greatest number in the long run," became a basic principle of conservation. In other cases, Muir in particular hoped to preserve natural areas from any immediate or eventual use. *Preservation* meant taking land and forests permanently out of the grasp of public and private users alike, reserving them for the simple activities he loved: hiking, observing nature, and reflecting on its meaning.

Muir's devotion to preservation became less flexible as he grew older, and he became increasingly dismayed at the rapid consumption of natural resources in areas he considered pristine. His preservation ideals came into conflict with the use-oriented conservation of the period. This conflict was not a simple one, since Muir counted among his friends those who advocated conservation, and since he, too, recognized the need for utility of resources. Yet, Muir chose certain fights and refused to back down.

Hetch-Hetchy Valley became the most famous battle between preservation and conservation interests. By about 1908, the city of San Francisco needed another source of water for its growing population, and city leaders turned to the possibility of damming Hetch-Hetchy Valley, not far from Yosemite. Despite end-

Plaques on the O'Shaughnessy Dam, which creates the Hetch-Hetchy Reservoir, Yosemite National Park, California. (Galen Rowell/Corbis)

less comparisons to Yosemite and the unique natural wonders of the valley, Hetch-Hetchy disappeared behind a dam in 1914. In the end, some believed Muir had lost not only the battle for the valley in particular, but also for preservation in general. The loss might have left Muir bitter, but he chose to mend the friendships strained by the conflict. He learned to hope that preservationists might gain from this one setback in order to succeed in future challenges.

The issue of protecting wild animal species blurred the distinction between conservation and preservation. According to Pinchot's view, hunters envisioned animals as a resource that needed management and conservation. A wider segment of society saw animals as part of the natural landscapes that Muir included in plans for preservation. From both perspectives, protection became the primary goal. From there, groups would argue for decades over what purpose that protection would ultimately serve.

People began to establish wildlife protection organizations in the late nineteenth and early twentieth centuries. They increasingly recognized that certain wild species, especially the abundant symbols of the American West, were in danger of disappearing. Others had already disappeared. The American bison went to the brink of extinction in the 1880s after decades of market hunting and the killing of these animals for only their hides or even simply for the thrill. The passenger pigeon, once the most abundant bird species in North America, went extinct in the wild by 1890. The last individual of the species died in the Cincinnati Zoo in 1914. These examples prompted people to worry that, without protection, other species would be lost due to human activities.

The rise of sporthunting also contributed to interest in conserving particular species, if only to ensure abundant game for future hunters. A growing appreciation for nature as recreation and as a source of social and moral strength further increased interest in wildlife conservation. Founded in 1888 by Theodore Roosevelt and George Bird Grinnell, the Boone and Crockett Club advanced the sportsmen's concern for a steady supply of big game. Roosevelt's commitment to protecting wildlife for future generations reflected his long involvement as a sporthunter. The emergence of sporthunting in the years after the Civil War marked a slow transition, almost unrecognizable as a threat to game species in the West. In a few eastern states, sporthunters organized game protection societies, and the regular publication of sporting journals also began in the late 1860s. In each of these forums, organizers and writers preached a certain amount of moderation in hunting. Grinnell was particularly outspoken against the hunting of bird species for profit on an open market. Based on experience elsewhere in the United States, a few early conservationists realized the most serious threat to game was sporthunting. If human hunting could be moderated, nature would find its own balance as it had in generations past, or at least that was what these conservationist leaders assumed.

The objectives of the Boone and Crockett Club included, first and foremost, the promotion of hunting and travel. In addition, the organization would "work for the preservation of the large game of this country, and, so far as possible, to further legislation for that purpose, and to assist in enforcing the existing laws" (quoted in Trefethen, *Crusade for Wildlife*, 356). These legislative aims served the cause of improving game conditions for the current and future generations of hunters. The next objective took a longer view and promoted natural history studies, including observation of animal habits. The club became a scientific institution, even if somewhat narrowly focused on game animals.

The Boone and Crockett Club was among the earliest of a proliferation of groups concerned with outdoor recreation, including fishing, camping, hiking, nature study, and ornithology. All of these activities, and their corresponding local and national interest groups, contributed to the early conservation movement. Another early national organization was the American Ornithologists' Union in 1883. This group formed the Committee for the Protection of North American Birds the following year, led by ornithologist William Brewster. Grinnell matched these efforts when he launched the first Audubon Society in 1885.

William T. Hornaday, another naturalist who promoted the cause of wildlife conservation, derived his interests less from sporthunting and more from concern over the broader significance of species extinction. Hornaday began crusading in the 1880s, writing and publicizing about wildlife destruction across North America and around the world. Hornaday's writing included descriptive zoological work in the late nineteenth century exploration and adventure style, and he prepared a four-volume collection of American natural history that was reprinted throughout his lifetime. Hornaday helped to found the National Zoological Park in Washington, D.C. and served as the first director of the New York Zoological Society's park, the Bronx Zoo. The work that brought him greatest renown was his book, *Our Vanishing Wild Life*, published in 1913. The book was a call to action for hunters, a call to "drop arms." He exclaimed that current attitudes toward wildlife resulted in slaughter and cruelty as a result of greed and selfishness. The book precipitated the establishment by Hornaday and others of the Permanent Wild Life Protection Fund in 1917. As the world went to war, Hornaday insisted that Americans needed to protect the resources of nature from neglect and destruction.

For many years, Hornaday operated as a kind of lone radical. His voice was heard, but his particular vision did not appeal to the sporthunting core of game conservationists or the nature-loving masses who romanticized wilderness. He painted a picture of the current situation that was too bleak to inspire hope, and his vision for the future required unpleasant change. He was out of step with the optimistic reformers of his day because he sought broad, common-sense alterations in wildlife protection of the sort that, by even the late 1910s, were proving problematic. Unlike those who favored scientific study and narrow expertise in providing specific and targeted solutions, Hornaday hoped to see broad changes—changes that would radically restrict hunters by expanding game protection. Such changes would also cripple livestock operations that demanded liberal use of public land free from competition with large, wild populations.

Professionalization of Science

Debates between preservationists and conservationists as well as continued pressure from scientists to protect natural areas developed against a backdrop of professionalization in science. Researchers felt the pressure of finding a place for their work, not only in terms of their individual careers but also in the broader context of society. Chemistry and physics became enormously more influential in this period, mostly through the development of laboratory research that demonstrated a clear experimental ethos. Laboratory experiments allowed scientists complete control over the conditions they studied. Individual variables could be isolated and compared under minutely adjusted circumstances. Control over conditions also meant scientists could more completely quantify their results. They could also replicate results, in a sense establishing the proof of their findings.

As experimental sciences, chemistry and physics departed from the traditional methods of natural history and natural philosophy. The study of the natural environment lacked the level of control achieved in laboratories. Many life scientists felt compelled to make their work recognizably modern by injecting an experimental ethos and expanding their use of the laboratory. Physiology and development became the laboratory-based standards of science in biology. Often, field studies provided results that appeared anecdotal by comparison to the standards of the laboratory, even when field scientists responded by adopting more quantitative and experimental methods. As such, studies of the environment were transformed by the experimental ethos into science with a level of rigor unknown in the nineteenth century. At the same time, however, many observers saw the link to traditional natural history as a de facto lack of rigor.

While many scientists believed that natural history had become inadequate to answer the needs of a progressive society, naturalists continued to ask the basic questions of plant and animal life histories. They also attended to the more practical issues associated with the distribution and control of various species. This period of transition studies took different forms and appealed to a range of practical audiences, including museums, universities, government-sponsored collecting and surveying expeditions, public education, and natural history societies. Depending on whether they were primarily interested in botany or zoology, naturalists focused their efforts for the professionalization of their science differently. Zoologists initially paid greater attention to physiology and morphology, making laboratory studies a more significant component of their research. Botanists also turned to physiological studies that they could conduct in laboratories and greenhouses, but they still attempted to embrace much of the natural history tradition.

During the 1880s, organizations of zoologists recognized that the status of their science required modernization. Descriptions of species, even new species, carried none of the excitement of laboratory work, especially when compared with the controlled precision of chemistry and physics. Life scientists recognized that laboratory work could be done, however, by paying closer attention to the physiological processes of animals. The motivation for keeping up with chemistry and physics included competition for funds for the building and expansion of institutions, for the brightest young researchers, and for recognition in the arena of public opinion. The American Society of Naturalists began to create a new image for zoology, joining with the American Morphological Society and eventually changing its name to the American Society of

Zoologists. The Society of Naturalists of the Eastern United States also contributed members to this new society of zoologists, which gave greater attention
to laboratory work and physiology.

Botanists likewise recognized the potential benefits of adopting an experimental perspective. They expanded studies of physiology to match zoologists
and added laboratory research, but they also recognized the ongoing need for
field research on plant distributions. This research, however, deserved an
update. Botanists distinguished the growing interest in plant interactions from
traditional descriptive work. Borrowing Ernst Haeckel's term for the study of
organisms in their environment, which he originally imagined to be of greater
importance to zoologists, botanists made ecology the name for their studies of
plant interactions. For the next decade or so, some botanists suspected that both
the name and the studies were a fad that would eventually decline. A generation
of students and younger researchers, however, would use the term and develop
the science into a significant field in its own right. In 1913, zoologists created a
new division within the American Society of Zoologists for ecology.

Conclusion

Progressive optimism in the United States fueled the growth of conservation.
People turned to scientific expertise as a guide to improving society, which also
meant responding to scientists' concerns about the depletion of natural areas
and the extinction of species, possibly jeopardizing the economy of nature. In
addition, they saw that same expertise as a potential key to unlocking greater
economic wealth from natural resources. Detailed studies of nature could play a
significant role in determining what nature had to offer in the late nineteenth
century. With increasing bureaucratic zeal, the U.S. government set out to study
natural resources, led by naturalists and administrators like C. Hart Merriam and
Gifford Pinchot. The bureaus created in the process transformed into management agencies for the nation's fish, wildlife, forest, and geological resources.
Finally, a growing awareness of the degree to which humans might determine the
future availability of certain resources, based on cases of abundance and
scarcity, led to a conflict over conservation and preservation ideologies.

Botanists and zoologists thus began to remake their fields in the context of
a constant tension between traditional natural historical studies that collected
basic data about the environment and the demand for information about how to
reshape the environment to serve human needs. That tension raised questions
that stimulated researchers in new directions. By the 1910s and 1920s, con-
tentious questions about the role of science in providing expertise limited some

scientists' ability to conduct research without feeling the necessity of answering to larger social, political, and academic institutions.

Bibliographic Essay

A recent book examining the development of environmentalism in the twentieth century provides easy access to the ideals and politics of Americans: Hal K. Rothman, *Saving the Planet: The American Response to the Environment in the Twentieth Century* (Chicago: Ivan R. Dee, 2000). Anthony N. Penna provides a wealth of examples in a book that can serve as a useful reference on natural resource issues: *Nature's Bounty: Historical and Modern Environmental Perspectives* (Armonk, New York: M. E. Sharpe, 1999). More comprehensive, but also less recent is Joseph Petulla's *American Environmental History* (San Francisco: Boyd and Fraser, 1977). A series of articles by Frank N. Egerton provide an introduction to issues in the history of ecology. Many of the valuable suggestions in Egerton's work for further research remain unexplored. See Egerton, "A Bibliographical Guide to the History of General Ecology and Population Ecology" *History of Science* 15 (1977): 189–215; and Egerton, "Changing Concepts of the Balance of Nature" *Quarterly Review of Biology* 48:2 (June 1973): 322–350. Gregg Mitman's *The State of Nature: Ecology, Community, and American Social Thought, 1900–1950* (Chicago: University of Chicago, 1992) is one of the few book-length accounts of the history of ecology that provides excellent treatment of the science, and in this case focuses on the group of ecologists who developed the science at the University of Chicago. Stephen A. Forbes wrote "On Some Interactions of Organisms" *Bulletin of the Illinois State Laboratory of Natural History*, 1 (November 1880), 3–17. The classic paper by Forbes, "The Lake as a Microcosm" (1887), is reprinted in an excellent collection of ecological papers: Leslie A. Real and James H. Brown, eds., *Foundations of Ecology: Classic Papers with Commentaries* (Chicago: University of Chicago, 1991), 14–27. That collection also includes a valuable introduction and synthesis of early ecology by Sharon E. Kingsland, "Defining Ecology as a Science" (1–13). Kingsland's book, *Modeling Nature: Episodes in the History of Population Ecology*, 2nd ed., (Chicago: University of Chicago, 1995) is another important work that outlines the development of ecology as an historical science. Keith R. Benson summarizes the transition from nineteenth century natural history, especially zoology, to twentieth century life science in "From Museum Research to Laboratory Research: The Transformation of Natural History into Academic Biology" in Ronald Rainger, Keith R. Benson, and Jane Maienschein, eds., *The American Development of Biology* (Baltimore: Johns Hopkins University, 1988), 49–83. The transition in botany is described by Eugene Cittadino, "Ecology and the Professionalization of Botany in America, 1890–1905" *Studies in the History of Biology*, 4 (1980), 171–198. The best book on the development of American ornithology and the role of scientists in conservation is Mark V. Barrow's *A Passion for Birds: American Ornithology after Audubon* (Princeton: Princeton University, 1998). John F. Reiger provides a comprehensive look at the role of hunting in early conservation in *American Sportsmen and the Origins of Conservation*, rev. ed. (Norman: University of Oklahoma, 1986). Another classic history of sporthunting is James B. Trefethen, *Crusade for Wildlife: Highlights in Conservation Progress* (Harrisburg, Pennsylvania: Stackpole, 1961). Joseph E. Taylor, III, analyzes the environmental history of fish culture in *Making Salmon: An Environmental History of the Northwest Fisheries Crisis* (Seattle: University of Washington, 1999). A classic in both progressive era history and environmental history that deals with science and expertise, especially in relation to water reclamation but also touching on a wider range of conservation issues, is Samuel P. Hays's *Conservation and the Gospel of Efficiency: The Progressive Conservation Movement, 1890–1920* (Cambridge, Massachusetts: Harvard University, 1959). Written at about the same time as Hays's book, a classic in the history of science outlining the role of government in American science is A. Hunter Dupree's *Science and the Federal Government: A History of Policies and Activities* (Cambridge, Massachusetts: Harvard University, 1957). A useful early history of forestry appears in Darrell Havenor Smith's *The Forest Service* (Washington, D.C.: Brookings Institute, 1930). More recent accounts provide a greater synthesis of how policies evolved. An excellent volume was edited by Harold K. Steen: *Forest and Wildlife Science in America: A History* (Durham, North Carolina: Forest History Society, 1999). Vernon Bailey wrote "Destruction of Wolves and Coyotes: Results Obtained during 1907," *Bureau of Biological Survey Circular* 63 (April 29, 1908): 1–11.

4

Ecology and the Foundations of Environmental Science

The sweeping changes in ecology that took place between 1915 and 1940, particularly in the United States, might be considered a revolution in science. With a focus on new questions and the application of ecological knowledge to problems of broad economic and political scope, ecologists became infused with a sense of the novelty and importance of their work, which their nineteenth-century naturalist predecessors had seldom known. Scientists, historians, and the general public now look back on this period, prior to World War II, and see the emergence of a new profession.

Both intellectually and professionally, ecologists distinguished ecology from the sciences of botany, zoology, and natural history in general. Intellectually, they focused research around a shared range of basic issues. Ecologists developed new frameworks for studying natural communities that shed the outmoded methods of natural history. They turned to examinations of species groups that conveyed a shared physiology, and while laboratories could not encompass complex communities, comparisons between groups began to reveal surprisingly simple underlying principles. Even without the laboratory, ecology became an experimental science, largely distinct from observational and descriptive natural history.

Ecologists created new institutions, societies, and journals to distinguish their science professionally even further from its predecessors. Professional practices conveyed the promise of a field of experts trained to study and solve the emerging problems faced by a society with a growing inclination toward upsetting nature's balance. Experts could contribute to preserving natural areas by demonstrating the dangers of human intrusion in nature. Professionalization also meant something less practical but far more significant for many scientists. They conducted research within formal institutions devoted to ecology, met regularly as part of specialized organizations, and published their research in highly regarded journals. Professional opportunities transformed ecology. This transformation of ecology into a legitimate science can indeed be viewed as revolutionary, but such a view should be balanced by looking at the continuities in science and studies of the environment.

Plant and animal ecologists worked together in establishing their science intellectually and professionally, yet they failed to unite their studies in the early twentieth century. Plant ecology remained closely linked to botany and strengthened ties to forestry, agriculture, and range management. Animal ecology kept close connections with mammalogy while increasing its associations with predatory animal control. Animal ecologists also faced new challenges in the emerging realm of game management. Despite their common efforts to understand groups of species, the traditions and subject matter of plant communities on the one hand and animal communities on the other remained an obstacle to a complete integration of ecology. The two groups had limited interaction despite explicit attempts to bridge the gap. The lack of integrated plant and animal studies made it impossible to answer certain questions, creating a situation that persisted for decades.

Ecologists Establish Institutions for Research

Academic institutions created spaces for ecology as investigators, including both university professors and graduate students, expanded their questions about plants and animals into research programs that required ongoing institutional support. The core of American science had rested in the academic institutions of the East Coast since their founding in the eighteenth century. By the late nineteenth century, new universities had emerged across the country, many of which became the homes of departments devoted to studies of the environment. On the West Coast, a special institution, the Museum of Vertebrate Zoology, was created within the University of California at Berkeley.

The Museum of Vertebrate Zoology was the first center of ecological research on the West Coast. Because of its ongoing status as a training site for a new generation of ecologists and as a source of criticism of government administration of wildlife, it stood out as an important institution in studies of the environment. Joseph Grinnell (1877–1939), the museum's first director, provided pioneering leadership far beyond his immediate administrative duties. In 1907, Grinnell taught natural history at Throop Polytechnic Institute (now Cal Tech in Pasadena). That year, he met Annie Alexander, a wealthy amateur naturalist who hoped to establish a museum for collections of animal specimens. Alexander came to Throop looking for a naturalist to accompany her to Alaska. Grinnell, who had spent a year in Alaska himself while in college, provided valuable advice to Alexander, and helped ensure the success of her expedition by

Zoology Building at the University of California, Berkeley, home to the Museum of Vertebrate Zoology, founded in 1908 by Annie Alexander. (Michael S. Yamashita/Corbis)

Joseph Grinnell (1877–1939)

Joseph Grinnell was born in 1877, far from any town in what is now Oklahoma. His father was a physician serving the American Indian communities there until 1885, when the family moved to Pasadena, California. Grinnell's devotion to natural history took root in California, and that state became his life's work. In 1906, he married Hilda Wood, with whom he had a daughter and three sons. He was an experienced collector and naturalist, having established his reputation even as a high school student. Grinnell spent a year in Alaska while in college.

He possessed exactly the right combination of research and administrative ability to help create an institution for the study and preservation of the specimens he and others had collected. Grinnell accepted the position of director of what would become the most significant zoological research centers on the West Coast, and eventually on the North American continent. The guidelines and procedures he established for cataloging and preserving specimens became the model for other institutions. He also developed protocols for collectors, which included meticulous preparation of specimens, as well as for each collection location: maps, photographs of the habitat, vegetation samples, and detailed field notes describing conditions. Such thorough information from the outset made the collections of the Museum of Vertebrate Zoology an especially valuable resource for researchers.

Grinnell published over 550 articles, many of them in ornithological journals. He planned to write a more synthetic work about ecology, evolution, and natural history when he retired, but never had the opportunity. When he died at age 62, his university colleague Alden Miller assembled some of his philosophical reflections on the study of nature into a volume entitled *Joseph Grinnell's Philosophy of Nature: Selected Writings of a Western Naturalist.*

recommending that she take Joseph Dixon, one of his students. From that initial contact, Grinnell and Alexander became serious collaborators for the rest of their lives.

Alexander shared Grinnell's passion for natural history, field biology, and collecting animal specimens. Grinnell hoped to focus on California as a region of rich and unique biological complexity due to its latitude, humidity, and distribution of plant species. Alexander offered the University of California at Berkeley a sum of $7,000 annually for the operation of a museum and asked the university to erect a building to house the collections and staff. She selected Grinnell to be the museum's director. By working and corresponding closely with him, Alexander retained almost complete control over museum operations, even as she traveled around the world and continued collecting.

From April 1908 until his death in 1939, Grinnell was the museum's first and only director, and he never wavered from his commitment to the institution. He considered leaving only once, in 1915, when colleagues at the Biological Survey asked him to apply for a position there. He declined and soon after became highly critical of the Biological Survey's program to control predatory animals. Under Grinnell, the museum's primary mission included cataloging and preserving specimens collected by Alexander and the museum's own staff, as well as generously sharing its collections with researchers from other institutions. Alexander had rejected the possibility of placing her collections in other universities or academies because of the potential limits those institutions might place on access for researchers. Although the Museum of Vertebrate Zoology had virtually no public exhibits, its collections were completely accessible to scientists who wanted to make use of them. Grinnell and Alexander agreed that their collections would be important for future generations in providing facts about each species and evidence of the process of evolution.

In the Midwest, a number of universities established a diverse set of approaches to the emerging science of ecology. Ecological studies at the University of Chicago, for example, emerged within an ongoing debate over the importance of experimentation in the laboratory versus broader physiological studies that required knowledge of an organism's natural environment. Laboratory studies considered the internal, even mechanical factors of organismic functions, which seemed to break with the methods of natural history. Part of this debate grew out of the fact that a geographer had helped establish the departments of botany, zoology, and geology. The geographical influence on faculty members of those natural science departments in the 1890s led to an enduring interest in the role of the physical environment in studies of plants and animals. As new faculty arrived with more training in experimental methods and with research programs oriented toward the laboratory, questions about the durability of natural history emerged. By the early twentieth century, students and faculty influenced by geography had moved on to other institutions or retired, and the laboratory became the focus of scientific research.

At the University of Chicago and elsewhere, laboratory studies lent themselves to questions of heredity, variation, and evolution. As a result, the university's first graduate with a Ph.D. in ecology, Charles C. Adams (1873–1955), combined lab and fieldwork to conduct evolutionary studies of ecology. In a slightly different way, Victor Shelford (1877–1968) began by dividing his research interests as a graduate student between evolutionary and physiological questions. Some instructors encouraged him to seek more causal explanations and, as a consequence, he adopted certain physiological approaches to ecology. He did this, however, without rejecting the experimental work that was becoming increasingly associated with the study of evolution. Shelford, like Adams, became a pioneer in animal ecology by recombining approaches that had previously begun to diverge. Henry Cowles (1869–1839) also helped pioneer plant ecology at the University of Chicago.

Chicago did not possess a monopoly on ecological studies, however. Other Midwestern institutions became important locations for cutting-edge science. Elsewhere in Illinois, Shelford spent most of his career teaching in the twin cities of Champaign and Urbana at the University of Illinois. Along with his many students, he broke new ground in animal ecology and community ecology. In Nebraska and later in Minnesota, Frederic Clements (1874–1945) laid the foundation for plant ecology, quantitative ecological studies, and community ecology. In Madison, Wisconsin, Aldo Leopold became the first-ever professor of game management in 1933, within the Department of Agricultural Economics.

Organizing Ecology

In planning an organization that would foster professional interaction across academic institutions, ecologists in America noticed that work in botany and zoology very rarely intersected. They decided that a group should devote itself to making such intersections more frequent. They wanted summer events in which plant ecologists could meet animal ecologists to conduct research together. In Great Britain, members of the Central Committee for the Survey and Study of British Vegetation established the British Ecological Society (BES) in 1913, with little attention to animal ecology. By contrast, Illinois plant ecologist Henry Cowles believed a professional society should provide plant and animal ecologists opportunities to work together. An American organization should be more inclusive from the beginning.

Cowles joined over fifty other scientists who gathered for a meeting in December of 1915. Fifty more wrote in, pledging their support for the new society. They succeeded in creating the Ecological Society of America (ESA). Members of the ESA elected animal ecologist Victor Shelford to be the society's first president, and they scheduled their first regular meeting for December 1916. In the meantime, they planned four summer research trips, with additional cooperative investigations in the works. Members intended these projects to serve explicitly as points of contact between plant and animal ecologists. Joining them were so-called "applied ecologists," with interests in forestry, *entomology*, and agriculture.

In Great Britain, ecologists did not see the need to seek the same inclusiveness. The BES created the *Journal of Ecology*, which included a majority of articles dealing with vegetative succession, plant evolution, and comparisons of floral communities. This was hardly surprising, given the origin of that organization in the British Vegetation Committee. Arthur G. Tansley (1871–1955), who served as the society's first president, worked as a botanist and plant ecologist.

In the first issue, Tansley expressed the hope that the new society could encompass the wider scope of ecology that had emerged in the previous decade. He acknowledged the rapid growth of ecology across the biological sciences. He anticipated that the new journal would never suffer from a lack of material to fill the journal's pages, even if most of its articles dealt with plant ecology. In stating that the *Journal of Ecology* would sample widely, Tansley considered general principles, concepts, methods, and surveys of greatest significance. The journal would also welcome articles dealing with the fields of geography, geology, meteorology, climatology, plant physiology, plant anatomy, and *floristic phytogeography*. In addition to these standards of plant ecology, readers would also find reports of studies in the special branches of science that Tansley believed were making great strides, such as soil science.

Across the Atlantic, similar and even greater ambitions marked the creation of the ESA's own journal, titled simply *Ecology*. In the foreword to the first issue, the editors invited papers from across the biological sciences, writing, "pages are open to all who have material of ecological interest from whatever field of biology." Barrington Moore, who became president of the ESA in 1919, described ecology as the most recent in a three-stage development of the biological sciences over the previous sixty years. Darwin and his colleagues had amassed knowledge that required only the simplest of observational methods. Others followed this stage with more specialized studies that led to the creation of zoology and botany. Around 1920, according to Moore, ecology would integrate biological science once again as part of a third, "synthetic stage." Far from containing all of the elements to complete this synthesis, ecology would be a common ground for cooperation between zoologists, botanists, foresters, and meteorologists. No single researcher could be all of these at once, and yet an ecologist would strive to take a broad point of view and seek out the cooperation of workers in related fields.

As an example of the synthetic stage of ecology, Moore described the forester who needed to reestablish a forest on 10,000 acres of burned-over mountains. The forester must secure the help of a meteorologist, soil scientist, zoologist, and phytopathologist—a scientist who studies the diseases of trees. Thus the forester, who would be recognized as such by his training and profession, would be considered an ecologist by adopting this broader perspective on his problem.

Moore also claimed that ecology included virtually all biological studies that were not particularly specialized. Agricultural research, animal industry, forestry, and even geography and history could be seen as ecological. It became a kind of rallying cry, as Moore encouraged readers of the new journal to acknowledge their potential contributions. He believed that rather than remain zoologists, botanists, and foresters, with little understanding of one another's

Forest recovering after fire, California. (Massimo Mastrorillo/Corbis)

problems, readers should endeavor to become ecologists in the broad, inclusive sense of the term. Such an approach would give them a chance to stake their claim in the expanding domain of science. Moore himself claimed the largest proportion of the scientific professions by including virtually all of the biological specialties under the umbrella of ecology, and the new journal would be the primary voice for the field.

While ecologists established a few specialized journals expressly for the publication of ecological papers, a number of journals that otherwise focused on natural history, ornithology, entomology, zoology, and botany published some of the foundational papers in ecology. The earliest included the *Bulletin of the Illinois State Natural History Survey*, in which students and faculty from Chicago and the University of Illinois published numerous articles. Other ecological work appeared in the *Botanical Gazette*, the *Auk*, *Nature*, the *Bulletin of the Torrey Botanical Club*, and the *American Naturalist*.

The publications board of the ESA established *Ecological Monographs* in 1931 in order to reduce the dispersal of articles dealing with ecological topics across so many journals. *Ecological Monographs* included longer treatises on detailed research. It included work in both plant and animal ecology, as had *Ecology* since its inception. In contrast, by the mid-1920s, the *Journal of Ecology*, which had consisted primarily of articles on British plant ecology, saw an increase in submissions dealing with animal research. Scientific journals that endeavor to include broad fields often face the tension of how to provide balanced coverage of topics. In response to the growing literature, the British Society launched the *Journal of Animal Ecology* in 1932. The society proposed that this new journal would consolidate animal research and prepare for future growth in that specific area. Meanwhile, editors of the *Journal of Ecology* continued to publish comprehensive biological surveys of plants and animals, as well as papers on the nature of *biotic communities*.

The Succession of Communities

As a botany student at the University of Nebraska in Lincoln, Frederic Clements worked closely with Roscoe Pound (1870–1964) to tackle one of the most troublesome tensions of working in a state-funded university. Administrators and legislators regularly reminded researchers of the pressure from the public—mostly farmers who were also taxpayers and voters—to produce useful knowledge. In Nebraska, that meant botanical studies that would increase the productivity and profitability of agricultural practices on the Great Plains. Clements and Pound studied under Charles Bessey (1845–1915), a professor of botany who took his science to new levels of respectability among academic scientists and government agricultural administrators alike. Bessey assembled a community of botanists at the university that included precollege assistants, postgraduate researchers, and students at every level in between. He was a teacher, mentor, administrator, researcher, and conservationist. Clements grew up on the Great Plains and became one of Bessey's star pupils, regularly participating in the informal but highly organized seminar of botany at the University of Nebraska.

Clements, Pound, and other members of the Nebraska seminar developed cutting-edge methods for plant ecology even as they began to inventory the usefulness of plants in Nebraska while doing postgraduate research with Bessey. In 1892, they expanded this program into a statewide survey. In the process of completing their report on the botany of Nebraska, Clements and Pound studied the writings of European plant geographers. There, they found the outlines of quantitative methods that would transform their inventory of plants into a full-scale survey that described the landscape. These quantitative methods included the development of the *quadrat* system, where discrete units of a community, usually one-meter square, could be examined intimately and the results expanded statistically to describe much larger units for comparison.

From his first biogeographical surveys, Clements developed a concept known as *succession*, based on a set of assumptions drawn from three sources. First, his reading of European botanical research granted Clements insights into

Frederic Clements (1874–1945)

Frederic Clements had the good fortune of marrying a botanist, Edith Schwartz, who collaborated with him on scientific projects and managed most of the practical aspects of their lives. Together, Frederic and Edith Clements traveled throughout North America studying plant communities. While Frederic's name is linked to the theory of ecological succession, Edith is best remembered for her memoirs of their work and for her illustrations, which often appeared in his publications.

Born and raised in Nebraska, his work progressed from groundbreaking descriptions of plant communities in his home state to foundational explanations of methodology for the newly established science of plant ecology. Most significantly, however, Clements conceptualized ecological succession as the process by which communities originate, develop, and reach a stable equilibrium.

With the techniques for description and comparison developed as a student of botany in Nebraska, Clements began to formulate his theory of succession. He published *Plant Succession* in 1916. By then, he was a professor and chair of the Department of Botany at the University of Minnesota. Clements's succession work cemented his reputation as the leading plant ecologist in North America. His theory, however, received less than universal acceptance. In Europe, where scientists and land managers had learned to take a longer view of what appeared to them to be very stable natural communities, Clements's theory had little appeal.

Ecologists—even those who disagreed with his ideas—respected Clements's research methods and dedication. British ecologist Arthur Tansley was among Clements's harshest critics when it came to conceptual issues, but Tansley nevertheless admired the tenacity and intelligence with which Clements pursued fresh knowledge. To the very end of his life in 1945, Clements was known as a difficult man to like. Even his friends admitted that he could seem arrogant and aloof. His powerful personality, however, contributed to the boldness of his ideas. It was Edith, his wife, who admitted that she sometimes had to remind him to eat and sleep, otherwise he would work with an intensity that led to complete exhaustion.

the broader comparative significance of his descriptions. Second, and of even greater importance, Clements's own experience as a naturalist on the Great Plains enabled him to make original contributions regarding the botany of the grasslands. Finally, he adopted an approach that assumed groups of organisms could be studied and referred to as parts of a larger organism, an approach known as *organicism*.

The European plant geographers provided methodology and comparative data as well as inspiration to do quantitative work. From that work, Clements also became familiar with botanists' descriptions of the consistency of areas where they had examined the details of soil and climate for diverse groups of plants. Eugene Warming (1841–1924) in Denmark observed how plant species interacted with one another in larger units that he began to think of as communities. Darwinian struggles for existence within these communities provided a larger framework for examining the growth of plants of different species. Warming also considered how communities changed over time as certain species increased and others decreased in numbers.

On America's grasslands, Clements referred to those groups of plant species found together, which Warming and a few others called communities, as *formations*. He observed that climatic and soil conditions provided a consistent basis for predicting what plant species would be found in a given area. When fire, drought, or human activities such as plowing disrupted the native plant life, leaving only bare soil, succession would eventually and predictably return most or even all the members of a previous formation to its native state through a series of stages.

The succession of stages in the development of plant formations depended on a return to stable conditions free of further disruptions. The native species found prior to the disruption would not generally grow on the bare soil left behind. Alterations in levels of moisture or nutrients caused by the disruption made it virtually impossible for original plants to resume their place on the land. Instead, a variety of *pioneer plants* would inhabit the soil. These plants often tolerated lower levels of moisture or nutrients. By growing on the bare soil, they began to change the conditions of that soil again, retaining moisture and returning nutrients. Their presence made it possible for other plants to take root. Plants in the second stage might grow taller, shading the pioneer plants. They might also have deeper root systems, thus being able to reach additional sources of moisture and nutrients. The formation of plants in the pioneer stage would eventually give way through competition for sunlight, moisture, and nutrients to another formation. In time, a third stage might out-compete the second stage, and so on until the original formation was restored. Clements referred to the final formation that would result from a given process of succession as the *climax stage*. A different climax might exist for each unique set of soil and climatic conditions.

At the core of his theory of succession, Clements proposed that since all plants were organized into formations, these formations could be studied as if they were organisms themselves. This proposal coincided with Clements's adoption of a broader organicism. He often insisted that plant formations were actual organisms and should be studied as such. As a consequence, scientists could describe the changes in vegetational formations as growth, maturity, decline, and even replacement, just as they would describe an individual organism. Formations grow directionally, continuing toward a climax stage, which was determined by factors such as soil type and climate. Clements generally argued that knowing enough about the climatic conditions of a region, a plant ecologist could predict the eventual climax community that would exist there.

In Illinois, Henry Cowles took a different view of succession. Trained as a geologist, Cowles examined the relationship between plant formations and the underlying geological formations on which they developed. From his perspective, changes could occur much more rapidly than Clements suggested, and those changes would not always follow the same linear pattern. Cowles did not repeat Clements's observations on the Great Plains. Instead, he went to the shores of Lake Michigan in northern Indiana, where sand dunes developed at the intersection of land and water, where wind and waves reshaped plant formations on a regular basis.

Cowles believed that sand dunes provided an ideal situation to study plant formations. Because of his interest in the dynamic nature of these formations, he found rare opportunities for plant ecology on the ever-changing conditions of the dunes. Cowles viewed the science of ecology as the study of processes in nature rather than the study of particular species. Nature displayed its processes more starkly to Cowles on the shores of Lake Michigan, where constant weathering action repeatedly wiped out previous formations and allowed the processes to begin again. He studied the development of a plant formation, beginning from the bare sand of a newly formed dune that wiped out everything below and before it in space and time. From bare sand, he observed the appearance of pioneer plants that managed to hold their own on the shifting dune. In time, shallow soils developed where other plants could take root. Over the course of several years, if new sand did not sweep in and bury the succession of formations, soils would deepen and, eventually, shrubs and small trees might take root. Closer to the lake, succession rarely proceeded this far, but as one traveled away from the lake, larger and more stable formations could survive. As he hiked farther from the lake, Cowles described the impression of moving through time, from the pioneer to more mature stages and on to the climax stage.

Cowles observed the many interruptions in succession on the dunes, where a new dune could form over part or all of a later stage, disrupting the process. He did not commit succession to regular progressive stages as Clements did. In some places, succession might appear to reverse itself, while in others it could be accelerated. Cowles studied the dunes for precisely this reason. He wanted to observe as many changes as possible to gain a greater sense of their causes and effects. He studied not only plants but also conditions of light, heat, wind, soil, and water. He considered the effects of fires, topography, and animals as the factors that contributed to the alterations of plant formations. As a result, Cowles provided a more complex study of plant ecology, and it proved less useful to casual observers. Unless one planned to spend years examining the changes of a dynamic place, the sand dunes did not offer much guidance for how a grassland or forest might respond to changes created by human activities. Clements, on the other hand, had offered a much more elaborate account of the stability of plant formations and how they might be managed to benefit agricultural and forestry practices.

The Dust Bowl and Sweeping Changes on the Great Plains

Clements believed his research would prove useful to farmers, but he and other ecologists could not have suspected that their ideas would be used to explain natural disasters. By acknowledging certain plant formations as climax stages, they admitted that planting crops interrupted succession or produced a replacement of the natural climax. The significance of this disruption might vary from year to year, but in times of broader changes in such factors as rainfall, the result of that disruption would prove unpredictable and ultimately devastating. During the 1920s, several years of drought brought agriculture on the Great Plains to a standstill, and when the winds blew across the bare soil, the storms that followed covered much of the central and eastern portions of the continent with dust.

On the open range, trouble had been brewing for decades by the time Clements and his colleagues began describing the ecological structure of America's grasslands. During the 1920s, increased agricultural use combined with lower than average rainfall spelled disaster. When drought arrived in earnest, people experienced storms of dry wind that kicked up unprotected topsoil and carried dust for hundreds of miles. These storms, collectively, became known as the "dust bowl." Clements's theory of succession played a significant role in understanding what was happening. He generally believed that agricultural practices had disrupted the natural succession of the community. Tilling the soil and grazing livestock removed the natural climax community of grasses. For those who believed in the inevitable progress of human society and technology, the theory of succession offered hope that, by understanding the natural process, humans could intervene and improve upon nature. Most ecologists recognized that the argument would take time and further study to resolve. In the meantime, there were no clear solutions that could stop the blowing topsoil.

Because of their new ecological knowledge of grassland communities, ecologists might have enjoyed a new level of authority in offering advice. Land administrators and agricultural officials turned to ecologists as land-use advisers to an entire nation. However, although Clements, Cowles, and others gained significant authority in the years leading up to this natural disaster, they were still a long way from being able to advise the nation on land management. The state of their knowledge combined with the extraordinary demands of a society that wanted its problems solved quickly forced ecologists to respond in one of two

As a result of poor land management and severe drought, spring winds carried off the topsoil of a large area in the southern Great Plains during the Great Depression. This 1936 photo shows the destitution of farm families threatened by the clouds of dust. The Great Plains Drought Area Committee was created to investigate the conditions that brought on the Dust Bowl. (Library of Congress)

ways. Some chose to retreat from situations that required them to offer practical advice, while others offered tentative advice that often proved to be less than adequate in practice.

Preserving Places

The government agencies that worked most closely on natural resource management during this period included the Bureau of Biological Survey, the Forest

Stephen Mather stands among a pile of antlers. He was director of the National Park Service from 1917–1929. (Corbis)

Service, and the National Park Service. At times, the goals of these agencies diverged dramatically and disagreements became significant both for science and public policy. The Biological Survey and the Forest Service emerged out of nineteenth-century administrative developments in the federal government. The Park Service was created in 1916 after vigorous lobbying by preservationists like John Muir and Stephen Mather (1867–1930). By then, the government had established several parks with only vague plans to manage those lands, including Yosemite Valley, Yellowstone, and the Grand Canyon. The appearance of a new agency to administer these substantial holdings became a source of controversy.

From the outset, the Park Service would not be directed like the Forest Service or even the Biological Survey, with a constant eye on resources. Mather, a businessman from the East Coast, became the Park Service's first director and actively supported protection of wild areas for tourists to see and enjoy. Inviting visitors from the East to see the uniquely American western landscape became his primary goal, and he used every possible incentive to coax people to travel across the country. He worked with railroad executives to encourage the expansion of rail lines that would run near national parks, and he hosted journalists to come and write about the natural wonders of America. Initially, Mather was more concerned about demonstrating the tourist value of the parks and he was not particularly interested in curtailing the utilization of resources administered by the Forest Service or in the efforts of the Biological Survey to exterminate predatory animals. However, when the activities of those other agencies conflicted with opportunities to boost tourism on or near his national parks, he launched massive public relations campaigns to promote the expansion of parks and to encourage preservation.

Mather's motivations for preservation occasionally intersected with the concerns of ecologists, but not because he shared their interest in understanding natural communities. The Park Service wanted a consistent image of a wild America to promote throughout its parks, including wild animals and serene forests. Mather himself regularly publicized the "balance of nature" that existed in the absence of human interference. By preserving the parks from such interference, he believed visitors could observe and enjoy nature in its true state. Ironically, building railroads to carry thousands of people to carefully preserved lands presented no contradiction for Mather. Others found preservation a more complex and potentially troubling proposition.

As an ecologist, Victor Shelford argued that complete ecological understanding of plant and animal communities required first that those communities be kept safe from the ravages of human interference. As the first president of the Ecological Society of America, he held a vision that included a firm commitment to nature preservation for purposes of scientific study. He was not alone in acknowledging that human activities across North America had drastically changed the landscape over broad regions, and he clearly expected his colleagues to join him in this call for preservation. Yet an active role in preservation suggested a link to political activism as exemplified by John Muir and the Sierra Club. The majority of ecologists, seeing the fragility of their new scientific and professional claims, opposed uniform participation in activities that would be viewed as nonacademic or overtly political. They wanted to focus on field research that would answer basic questions about the relationships among plants or between animal species.

In 1917, ecologists found a way to compromise. A twenty-five-member committee of ESA members planned to list all of the preserved and preservable natural areas in North America. The details of such a list would be a fundamental contribution to science, and would thus enhance ecologists' respectability among scientists. The list would also follow Shelford's mission to focus attention on preservation by highlighting the value of specific areas. Committee members admitted that their task was made inestimably more difficult due to the lack of any comprehensive guide to natural areas. They decided to divide the work by assigning ecologists to every state and province in North America and collecting data simultaneously and without delay across the continent.

Victor Shelford (1877–1968)

Victor Shelford was born in New York state in 1877, and grew up about a mile north of the Pennsylvania border in the town of Chemung. He attended small country schools and chose to continue his education by attending a teacher's college even before he had a high school diploma. In fact, because he took classes wherever he could, he did not officially graduate until he enrolled at the high school in the nearby town of Waverly. He was almost twenty-two years old when he finally got his diploma.

In the summer of 1903, after receiving his bachelor's degree from the University of Chicago, Shelford took the first of many trips across the country. He always made observing and collecting information about ecological communities an important part of these trips. On a trip to Europe in 1907, the only difference in his scientific agenda was that he had brought along his new wife, Mabel. While it was their honeymoon, he still took advantage of the opportunity to visit museums in order to examine their tiger beetle collections. Mabel was a schoolteacher and a keen participant and collaborator in ecological research.

Shelford did his graduate work at the University of Chicago, studying with professors who introduced him to the science of ecology. After receiving his Ph.D. in 1907, Shelford was hired by the university to teach field zoology and develop a natural history curriculum. During the summer of 1914, while he was doing research at the marine laboratory in Friday Harbor, Washington, Shelford secured a permanent faculty position at the University of Illinois in Champaign-Urbana, which became his home for the remainder of his career.

As the first president of the Ecological Society of America in 1915, Shelford also led efforts to preserve natural areas for educational, scientific, and recreational use. He believed scientists should play a leading role in preservation. When Shelford died at the age of 91, he was perhaps the most highly respected ecologist the field had seen. He established the standard, through his instruction as a teacher and by example as a researcher, for long-term, quantitative studies that integrated field with laboratory, aquatic with terrestrial, plant with animal, and observation with experiment.

By 1924, seventy-five members of the expanding committee on preservation had compiled information for the purpose of maintaining biotic balance in existing preserves. They had also worked to find an organization in the local area of each preserve that would work in the interest of pure science. These organizations would not be subject to pressure from agricultural, mining, fishing, or forestry interests, and would therefore help to maintain the balance of nature in perpetuity. Ideally, these groups would serve as part of a state academy or natural history society. The success of this committee in producing *The Naturalist's Guide to the Americas* (1926) testified both to Shelford's determination and to the relative scarcity of such areas.

Although preservation efforts produced concrete results, ecologists rarely saw those results as the ideological core of their new profession. They faced a tension over how ecologists would serve as an objective source of natural knowledge and, at the same time, remain committed to preserving nature. It became a struggle to establish and maintain academic respectability. Much of the struggle in the early years of the Ecological Society of America revolved around the efforts of a few influential ecologists to make the new, professional society into an agent for the promotion of nature preservation.

Shelford believed that ecologists' professional mission should be intimately connected to preservation efforts, but many of his colleagues disagreed with his perspective. Thus, despite the fact that Shelford served as the society's first president and worked tirelessly to promote nature preservation by creating special committees and reports, his efforts produced little fruit within academic ecology. Nature preservation did not emerge in this era with a scientific mandate, nor was it a clear popular priority. Rather, efforts of scientists and the attention of the public turned to more specific debates over particular species.

Protecting Species

Preserving natural places created conflicts with human activities at the borders of designated parks, forests, and valleys. Government agencies often got caught in the middle of these conflicts, with mandates to conserve natural resources, provide for nature preservation, and enhance economic opportunities on public lands. Scientists played a role in attempting to mediate these conflicts by providing expertise based on studies of ecology and natural history. Clear scientific advice rarely emerged in the midst of complex questions of politics, economics, and the more intractable demands of a society that wanted to preserve nature as well as prosper by harvesting its resources.

Efforts to protect wildlife in the early twentieth century involved a mingling of conservationist and preservationist motivations. President Theodore Roosevelt saw many animal species as game, and as a sportsman he valued those species above all others. Without protection on federal lands, he feared game species would become too scarce for the sport of future generations. Preservationists shared an appreciation for many animal species, regardless of their status as game or nongame animals, hoping to protect them for the enjoyment of future generations.

In response to the broader support for preservation and combined with his own impulses as a big game hunter, Roosevelt began establishing game preserves. Without any model for how to protect wildlife on federal lands, he simply called for restrictions on forest lands administered by the Department of Agriculture. This move ignored the fact that individual states had jurisdiction over hunting and that federal preserves might conflict with the states' rights. However, game preserves were an experiment, and the products of a few of these experiments demonstrated the shortcomings of wildlife protection. Preservationists, hunters, state game departments, and land use managers all began to challenge the existence of federal preserves.

Most notable among western big game preserves, Yellowstone National Park and the Grand Canyon National Game Preserve faced apparent crisis situations by the late 1910s and 1920s. Around Yellowstone, elk could not find sufficient food to support the growing population in the preserve. North of the park, elk began grazing on privately owned rangeland, much to the displeasure of ranchers. The animals broke down fences to eat hay that ranchers had put up for their cattle. South of the park, elk hungrily stalked into the town of Jackson Hole, Wyoming. The National Park Service began feeding the elk to prevent further damage to private land and property. Discussions of the other alternative, killing the excess animals, brought widespread criticism from all sides. Hunters would rather have sporting opportunities within the preserve than stand at the boundaries firing at animals as they exited. Killing elk within or just outside the park appalled preservationists who would rather expand the preserve to accommodate the elk. That solution would mean turning private ranch land over to the Park Service. Ranchers who owned land and made their living around the preserve found this suggestion equally appalling. Others hoped that Yellowstone could export live elk to parts of the country where they had already disappeared and herds might be reestablished. Zoos would appreciate a ready supply of the beasts.

On the Kaibab Plateau, just north of the Grand Canyon, government officials, scientists, and the general public observed the results of another of the earliest wildlife protection efforts. The case of the Kaibab deer became well known to every conservationist, nature lover, hunter, and ecologist. The problem that arose resulted from the perception that the deer population had grown too large after Theodore Roosevelt declared the Kaibab Plateau a game preserve twenty years earlier. In 1906, Roosevelt placed the deer in the care of the Forest Service, which had administered a forest reserve there since the 1890s. Between 1906 and the early 1920s, local foresters, ranchers, and hunters watched the deer numbers increase. They later estimated that the herd included about 4,000 animals at the turn of the century. By 1924, some ranchers estimated that the herd had grown to 100,000 deer, and a few local foresters agreed. Others, especially those who had no interest in seeing cattle or sheep in the game preserve, estimated the deer numbers to be closer to 25,000.

Yellowstone National Park: Bison herd roaming the valley of Yellowstone National Park. (National Park Service)

Whatever the actual count, in some parts of the preserve, forage became scarce by the early 1920s. Although the scarcity of vegetation resulted from multiple factors, including drought and livestock grazing, many observers feared that the deer were suffering, and some suspected that game protection efforts caused the rapid increase of their population, creating this fate for the deer. People began to wonder what could be done to halt the increase of deer and the damage to vegetation. Proposals included killing deer or capturing and transporting them to other preserves. A few people even hypothesized that since ranchers, bounty hunters, and government trappers had pursued wolves and mountain lions ever since the preserve was established, the loss of predators had upset the delicate balance of nature.

While the debate raged over what to do next, ecologists began to examine the conditions on the Kaibab Plateau. Their studies suggested that the deer population, while probably higher than its ideal *carrying capacity*, had reached a maximum of about 30,000 deer, and by 1930, it was around 20,000. So while extreme exaggerations became widely publicized, the work of scientists, alongside state and federal administrators, worked out a system of hunting that would help to stabilize the deer and their community. However, this scientific work could not immediately quiet the controversy over the need for changes in deer management. As a result, misunderstandings and exaggerations transformed the situation into an infamous textbook case of how humans disrupted the balance of nature through their shortsighted actions.

Controversy in Ecology

By 1924, ecologists believed they knew enough about the mismanagement of federal wildlife programs to mount a campaign for reform. While the Ecological Society of America remained unprepared to enter controversies over government policies, the ecologists who chose to speak out at this time did so as members of the American Society of Mammalogists. The concerns of Joseph Grinnell, Joseph Dixon, Charles Adams, and others prompted the American Society of Mammalogists to form a Committee on Wild Life Sanctuaries, which included ecologists along with government biologists. They agreed to help the government formulate policies for the preservation of wildlife, including predatory animals. The committee would also recommend the location of preserves for game animals as well as for the animals the government had worked to exterminate since 1915. Adams chaired the committee and was joined by Dixon, Vernon Bailey, Edward Goldman, and Edmund Heller. The Biological Survey's systematic efforts to eliminate predatory species prompted concerns about the growing need for nature preserves. With Bailey and Goldman representing the government predatory animal control program, the committee quickly became a forum for discussion of ecological ideas, particularly the balance of nature.

The professional training of Biological Survey employees became an important feature of the predatory animal control debate. Bailey and Goldman lacked the academic credentials of their ecologically trained counterparts. They nevertheless undertook expansive research projects, surveying numerous species in large regions of North America. These men published their findings in Forest Service and Biological Survey circulars and in journals of natural history and mammalogy. They enjoyed a respected status among their academic peers. Bailey and Goldman were exceptional, however, since most Biological Survey employees worked at more practical tasks deemed important to the public good, such as serving as game wardens and trapping predatory animals, but this work had minimal scientific value. For the majority of Biological Survey workers, their primary duty was to protect agricultural enterprises in the most effective way possible, such as destroying wolves, coyotes, and other predatory animals.

The Biological Survey faced increasingly difficult questions regarding the role of predatory animals and the balance of nature from ecologists, and also from Mather with his plans to increase tourism in the national parks. Some ecologists believed the government was bending to the wishes of special interests, including the livestock industry and sporthunters. They were convinced that the government destroyed potentially valuable animals in deference to a few narrow special interest groups. Ecologists suggested that more research was needed to demonstrate the potential value of predatory animals. Grinnell offered the ecological view that predatory animals might actually provide a practical service in limiting rodent populations.

While many ecologists insisted the Biological Survey should study the life histories of predatory animals before exterminating them, Goldman attempted to convince his audience of the necessity of immediate and ongoing government control efforts. The government biologist pointed out that nature was complex and that complexity led to differences of opinion concerning predatory animals. Goldman had seen firsthand how difficult it would be to unravel the combined influences of predation, livestock, grazing, and protection from human hunting. He explained that his opinion differed from that of many others due to his deep understanding of the full information regarding the status of predatory mammals. He called for consideration of the problem in its complex and diverse settings.

Goldman's attitude toward the balance of nature signified an important shift in the thinking of many scientists and naturalists. He maintained that, although balance had long been considered a fundamental feature of nature, in the twentieth century, balance could only be considered hypothetical. The notion that species remained in balance, neither increasing nor decreasing, could not be supported, particularly when observing the many changes in population taking place as a result of the activities of humans. Goldman suggested that before the so-called discovery of North America, large carnivores, including what he called primitive human tribes, limited game animals across the continent. He asserted that Europeans bearing firearms violently overturned the balance of nature, and it could never be reestablished. By 1924, humans had confined game animals to the western mountains, particularly in parks and national forests. Predatory animals followed their prey and now threatened to destroy not only livestock but also the relatively few game animals that remained. For a variety of reasons, predatory animal control served the purposes of practical conservation. In Goldman's view, control provided a simple solution to an otherwise impossibly complex problem.

Practically oriented federal biologists in the Biological Survey sought to limit the damage to game and livestock done by predatory animals. At the same time, they believed they could limit the damage done by rodents through the use of poison. Goldman noted the greater effectiveness of human efforts in reducing these rodent populations, rather than remit control to natural enemies. He insisted that to allow predatory animals to persist, expecting them to control rodents, was to miss the point of civilization. In his view, civilized man had overturned the balance of nature by creating artificial conditions. For this reason, practical considerations demanded that humans assume effective control over wildlife everywhere. The Biological Survey could take responsibility for wildlife, predatory animals, and rodent populations, promoting the beneficial and destroying the harmful.

Ecologist Charles Adams challenged Goldman's argument that the balance of nature had disappeared when Europeans arrived in North America. Describing changes in the balance of nature since the arrival of Christopher Columbus in 1492, the ecologist stated that thousands of years earlier, nature maintained a relative balance of species and that American Indians had been a part of that relative balance. Recently, the changes begun by adventurers and trappers accumulated to the point that the former balance of pre-Columbian days was indeed gone forever. With the activities of humans in the last few centuries, Adams claimed, nature had begun to adjust toward a new balance. That new balance in North America included the establishment of a vast public domain. While wildlife thrived on that public domain, the future value of this wilderness was being spent rapidly. The federal agencies that administered public land should rely on ecological knowledge in establishing management plans. Adams suggested a major role for scientists who would formulate definite ideals and policies that would guide naturalists, conservationists, and policymakers alike.

From his general statement of the need for scientific management and far reaching conservation efforts, Adams went on to describe the particular status of predatory animals, which faced immense pressure from human intervention. The relationship between predatory animals and the balance of nature, according to Adams, suggested that conservation of wildlife necessarily depended upon a better human understanding of all animals. He also acknowledged that because predatory animals presumably played an important role in preventing overpopulation of prey species, they were worthy of study. Furthermore, scientists required wild specimens to understand that role in more detail.

Adams believed the Biological Survey needed to decide how significant samples of predatory animals would be conserved before continuing with its campaign to exterminate them. Conservation depended upon a basic understanding of the relationship between the land and its people. The ecologist recommended that studies of predatory animals be conducted on urban lands, agricultural lands, and wild lands in the national and state parks and forest systems. On agricultural lands in particular, Adams pointed out that the Biological Survey and livestock owners had failed to support such studies in the past, but had spent millions of dollars on the extermination of predatory animals. He suggested that some of this money should be diverted into scientific studies. This comment

Charles C. Adams (1873–1955)

In the early twentieth century, hardly any naturalist could be referred to consistently as an "ecologist." Charles C. Adams may be the most likely exception to this generality. Adams was, in fact, the first person to earn a Ph.D. in ecology, rather than some related field like botany or zoology as did most of his colleagues. Despite that distinction, and the enormous amount of well respected research he did in his career, this first ecologist faced many obstacles throughout his career.

He was born in 1873, and like many boys in rural America in those days, he grew up surrounded by questions about the natural world. Adams began publishing his observations of natural history in his home state of Illinois while he was a college student. He eventually published over 150 works, including a valuable reference work on early ecological literature.

Adams did not follow the typical academic career. He earned a master's degree from Harvard before heading back to Illinois, where he finished his doctorate at the University of Chicago in 1908. Along the way, he worked for over three years in natural history museums at the Universities of Michigan and Cincinnati. This experience convinced Adams that museums could and should serve as vital links between field research and education about nature. Most academic scientists believed universities should serve exclusively in that role, giving little thought to museums and other public institutions. Adams, however, proved to be unconventional in his support of museums, and his unconventional suggestions did not end there. He taught at the University of Illinois until 1914, but was never offered a permanent position there. As a result, he moved to New York as a professor of forest zoology. While there, he established and became the director of the Roosevelt Wild Life Research Experiment Station. He held that position until being named director of the New York State Museum, which brought his career full circle back to museums.

In conducting research and promoting conservation of wildlife, Adams attracted the attention of high-ranking officials in the National Park Service. At one point, he requested documents about wildlife policies—documents that the agency had previously released to journalists. When one official withheld the documents, others quickly insisted that the material be turned over to Adams in order to avoid public criticism, which Adams would be likely to invoke. In correspondence and publications, he frequently offended and insulted his colleagues, seemingly without realizing the offense. Those who found his manner disagreeable responded at times by thwarting his advancement in academic positions.

struck raw nerves among officials in the Biological Survey, but Adams stopped short of accusing the government of using taxpayer dollars to protect the investment of livestock owners from predatory animals rather than conducting research in the broader interest of the public.

In order to preserve wildlife, including deer, elk, and even cougars, Adams argued that forward-thinking conservationists would need to establish more large refuges like Yellowstone and the Kaibab. Rather than mere game preserves, these would serve scientific, educational, recreational, and economic purposes, and would necessarily include large predators. On the Kaibab, for example, a few observers had begun to blame the elimination of predatory animals as a cause for the deer overpopulation. Attempts to include predatory animals in these refuges, Adams cautioned, would require an understanding of the relationship between predatory animals and prey. Along the boundaries of these preserves, administrators should expect some overflow of mountain lions, coyotes, and wolves. He believed proper control would be needed in these areas, but only as a result of careful scientific studies.

Edward Goldman and Charles Adams represented the range, if not the entire spectrum, of attitudes toward predatory animals and the balance of nature. Ecologists interested in studying the life histories of all mammal species hoped to preserve, at minimum, samples of predatory animals in order to examine their food habits. Even more important, many hoped to study these food habits, especially in regards to rodents and game as prey. Of course, it only occurred to ecologists that predator-prey research might be useful in wildlife conservation after questions of game management arose. Scientists needed to invent methods (such as game survey techniques) and define concepts (such as carrying capacity) that might begin to answer rapidly mounting questions in this unique context. Those questions involved relationships between deer, mountain lions, coyotes, livestock, and other factors in preserves. Because these questions fascinated people with widely ranging interests and practical concerns, ecologists were not allowed the luxury of conducting their studies in seclusion. They felt the pressure of providing solutions to practical problems using the knowledge developed in ecological studies.

New Principles in Animal Ecology

While Frederic Clements had given botanists and range managers alike new principles to guide their investigations and planning, animal ecologists continued to struggle amid the perceived complexity of their subject. Compared to plants, animals seemed difficult to study: they moved around, they hid from plain view, and they pursued many of their activities in the dark of night. Moreover, the complexity of their interactions defied easy explanation, as exemplified in the debates over predators. In 1927, a young British ecologist synthesized many of the most complex problems in surprisingly simple terms. Charles Elton wrote *Animal Ecology* in a period of less than six months, introducing concepts like *food web* and the *pyramid of numbers*. These terms proved extremely useful to ecologists and zoologists alike by establishing underlying principles for the comparison of different kinds of communities, and even for bringing similar questions to both plant and animal ecology.

Elton considered ecology an extension of natural history, but his models for natural interactions included more explicit economic terms. He believed ecologists would have to abandon vague notions and develop firmer concepts. Instead of a balance of nature that implied constant populations persisting over time, he sought an understanding of the ceaseless fluctuations of species. He referred to plants, which converted raw materials like sunlight, soil, water, and carbon dioxide into food, as *producers*. The animals that ate those plants he called *consumers*. Producers and consumers perpetually increased and decreased in numbers as they responded to changes in the broader community.

Elton's pyramid of numbers reflected his economic view. The pyramid consisted of a broad base that included all producers. Since plant populations included the largest numbers in any community, those producers could support a wide variety of consumers. He conceived of the consumers in multiple levels above the producers. The animals that ate plants, the herbivores, were less numerous than the plants themselves, but more numerous than the carnivores that fed on herbivores. Many communities had one or two additional levels of consumers above these, carnivores that eat other carnivores, taking the pyramid upward in ever smaller numbers. The highest level of consumers, the large predatory animals and birds of prey, would be the fewest.

The pyramid of numbers demonstrated three other principles of ecology: *niches*, size of food, and food webs. Elton built these principles on the work of other ecologists, but combined them in a way that showed how ecology, as scientific natural history, would examine them all as closely linked. Niches, first proposed by Joseph Grinnell over a decade earlier, were the roles that particular species played within the community. Where a species lived and what it did were only slightly less significant than what it ate and what ate it in defining its niche. Within Elton's economic approach, niches were the locations where species performed key industries, providing food for others and finding food for themselves. Another factor in determining a species' niche involved the size of food eaten by that species. Animals tended to prey upon others of a smaller size. Elton admitted that this observation seemed obvious and had many exceptions, but he presented the size of food as a range. Within the range of possibilities, certain optimum choices could be made. Assuming it could catch and kill them, a fox might eat tiny insects or even a deer, but its optimum food included mice on the small end of the range and rabbits on the large end. By proposing this principle, Elton could at once claim that spiders do not catch elephants in their webs, and more significantly that the limits of food size could help ecologists to understand the more likely interactions among species.

The most familiar among Elton's principles, food webs, derived from the simpler notion of *food chains*. He believed ecologists could study the flow of energy contained in food from one individual to another. This flow could be traced through the levels of the pyramid of numbers, but it could also be traced from species to species, including from plant to animal. In this view, each species within a given community formed a link in the chain. Elton quickly pointed out that most animals consumed more than one other species for food. Many animals subsisted on plants as well as animals from multiple levels in the pyramid of numbers. The ecologist proposed that the relationships could best be represented as a web of interlinked chains. His sketches of food webs were complex, reflecting the many relationships that ecologists interested in the structure of communities must study. This and Elton's other principles of animal ecology established relatively simple underlying notions of how ecologists could understand otherwise bewilderingly complex communities of organisms.

Managing Nature

While ecologists organized their profession and established new concepts and methods for their science, the potential utility of ecology also expanded. For decades, Americans heard the promise of scientific expertise and rational management touted from the highest levels of government. President Theodore Roosevelt and his chief forester, Gifford Pinchot, had based an entire federal program of land and resource management on the premise that insights from science and technology would improve and transform nature from a wasteful wilderness to an efficient source of raw goods. With the support of the entire federal government, the Forest Service and its parent agency the Department of Agriculture began an idealized assault on resources, especially in the western United States.

Western forests in the Rocky Mountain states and near the Pacific Coast seemed vast and inexhaustible. The individual trees in many of these forests were far too large to be processed by most nineteenth-century lumbering technologies. Especially in the Pacific Northwest, Douglas firs, Sitka spruce, red cedar, and the various pines and firs grew to enormous sizes on the moist mountain slopes. Besides their favorable growing conditions, these trees were very old—500 years old and more. In most mountainous regions, forest fires served to clear the undergrowth and smaller trees beneath the canopy of the mature forest, improving rather than threatening the growing conditions for the old giants. Lumber companies soon adapted new technology to these conditions.

Tourists examine the remains of an ancient tree in California. Untouched by human activity, many trees can live over 1,000 years. There are few old-growth forests left in the United States, and their preservation has been the object of heated debate and civil disobedience actions against logging companies and the U.S. Forest Service. (National Archives)

In addition to engines that could drive saws and mills, which reduced the need for human labor and proximity to water power, transportation improved accessibility to forests. Caterpillar tractors could reach felled trees and haul them out with relative ease. In the 1930s, lumber operations included road building so that trucks could drive in and out of the forest, transporting loads of timber to increasingly distant destinations. Areas once considered too remote to provide a profitable return on the investment became cost effective for forestry. Railroads across the West had already opened markets in all regions of the country to the fine, tall, straight timber of America's mountain forests.

As part of this broader utilization of resources, the federal government also transferred increasing amounts of land in the South to private ownership. In contrast to land managed by the government, these forests then declined rapidly. New technology enabled landowners to harvest more trees more expeditiously than ever before, making profits more immediate and reducing the incentives for conserving private forests. While ecologists developed a greater understanding of natural communities in one part of the country, social and political changes elsewhere made the goals of applying their knowledge problematic. Forest management increasingly focused on raising profits by preventing fire and disease.

In the late 1920s, Aldo Leopold abandoned a career in forestry to begin studying wildlife populations in order to attempt to manage them more like forests. Trained at Yale in the modern methods of forest management, Leopold realized that forests were more than just trees. His experiences on federal lands in New Mexico taught him that interactions among ranchers and loggers figured into an equation that also included a wide diversity of natural features. Reading about wildlife protection in the works of conservationist William Hornaday inspired him to take a more active interest in the wild species of the forests. He also recognized the toll of soil erosion in areas where human activities led to overgrazing by domestic animals. These larger changes in nature prompted Leopold to leave the southwest and eventually to leave the Forest Service but to retain the management goals of that agency. He embarked on an unprecedented game survey of the Midwestern states, a project that expanded into a new field of game management.

A lifelong hunter, raised on the banks of the Mississippi River in the late nineteenth century, Leopold turned his enthusiasm for wild game into a full-time livelihood. He received funding from an organization of gun and ammunition companies and began collecting information on various birds and mammals, primarily species hunted for sport. Traveling across the Midwest from his home in Madison, Wisconsin, he spoke with state officials, hunters, and naturalists to compile a record of game populations. He also read the work of animal ecologists like Charles Elton, Victor Shelford, and Charles Adams, but his forestry background led him to approach ecological questions differently. Rather than study populations and fret over the need to preserve them in some difficult-to-define natural state, Leopold fixed on the notion of management. Game, like trees, might yield to efforts to control conditions that would promote growth and reproduction. Hunters stood to benefit most from the implementation of measures that could increase game production. As naturalists had learned from cases like the Kaibab, more protection could lead to unpredictable and unwanted results. Leopold discussed these potential advantages in *Game Management* (1933), the first textbook in this field.

Leopold popularized the field of game management (sometimes referred to as wildlife management) with reference to cases like the Kaibab. He especially emphasized how the removal of predators had left the deer with few enemies and may have contributed to what he began to call an *irruption*. Leopold believed an irruption was part of a population cycle where numbers increase rapidly in response to some change in conditions, ultimately resulting in a shortage of food and the starvation of large numbers of animals. He further suggested that irruptions could reduce the carrying capacity of an area as the overpopulated species deplete available food for generations to follow. More detailed ecological studies have since demonstrated that this idea of Leopold's relied too heavily on simple logic and not enough on actual data. His ecological colleagues did much more to study and understand the fluctuations of actual populations. Leopold eventually became best known for his more philosophical musings on the importance of wilderness and the proper means of maintaining and appreciating nature as part of a *land ethic*. The way he introduced management goals into the field of wildlife conservation remains a much debated component of how ecology revolutionized studies of the environment during this period.

Conclusion

In the 1920s and 1930s, ecology emerged with many of the fundamental features of environmental science. Scientists had developed important theoretical concepts for describing natural phenomena, and they were raising new questions about how the balance of nature could be explained or challenged. Although it diverged significantly from nineteenth-century natural history, the changes in ecology during this period relied heavily on past assumptions and methods. Many of the new ideas in ecology derived from related fields of inquiry, where different motivations prompted novel investigations. Clements and Cowles developed notions of succession out of a careful study of botanical and geological conditions. On the Great Plains, land administrators and ranchers demanded that ecologists attempt to apply their ideas in ways that would enhance agricultural practices. Ecologists acquired new authority from their science in the midst of these demands in land use planning, preservation, wildlife protection, and forestry.

Bibliographic Essay

Sharon E. Kingsland's excellent article "Defining Ecology as a Science" appears in Leslie A. Real and James H. Brown, eds., *Foundations of Ecology: Classic Papers with Commentaries* (Chicago: University of Chicago, 1991), 1–13. Robert A. Croker has written a valuable biography of Shelford, which includes a broad view of both his scientific and preservationist activities: *Pioneer Ecologist: The Life and Work of Victor Ernest Shelford, 1877–1968* (Washington, D.C.: Smithsonian Institution, 1991). Wildlife protection is addressed broadly in Thomas R. Dunlap's *Saving America's Wildlife: Ecology and the American Mind, 1850–1990* (Princeton: Princeton University, 1988), and Donald Worster's

Nature's Economy: A History of Ecological Ideas, 2nd ed. (New York: Cambridge University, 1995). The cases of Yellowstone National Park and the Kaibab deer are detailed in James A. Pritchard's *Preserving Yellowstone's Natural Conditions: Science and the Perception of Nature* (Lincoln: University of Nebraska, 1999) and Christian C. Young's *In the Absence of Predators: Conservation and Controversy on the Kaibab Plateau* (Lincoln: University of Nebraska, 2002), respectively. Ronald C. Tobey details much of Clements's work and the history of plant ecology in *Saving the Prairies: The Life Cycle of the Founding School of American Plant Ecology, 1895–1955* (Berkeley: University of California, 1981). Joel B. Hagen provides an excellent overview of the history of ecology: *An Entangled Bank: The Origins of Ecosystem Ecology* (New Brunswick, New Jersey: Rutgers University, 1992). A more comprehensive but less readable account is Robert P. McIntosh's *The Background of Ecology: Concept and Theory* (New York: Cambridge University, 1985). Histories of the Forest Service and National Park Service abound, but some of the best and most readable sources include Alfred Runte's *National Parks: The American Experience*, 2nd ed. (Lincoln: University of Nebraska, 1987); Robert Shankland's *Steve Mather of the National Parks*, 3rd ed. (Alfred A. Knopf, 1970); and Harold K. Steen's *The U.S. Forest Service: A History* (Seattle: University of Washington, 1976). A fine biography of Annie Alexander is now available in Barbara R. Stein's *On Her Own Terms: Annie Montague Alexander and the Rise of Science in the American West* (Berkeley: University of California, 2001). Historians have written relatively little on the early years of the Ecological Society of America, but Sara Fairbank Tjossem describes the controversies mentioned here in "Preservation of Nature and Academic Respectability: Tensions in the Ecological Society of America, 1915–1979" (Ph.D. dissertation, Cornell University Press, 1994).

5

Radioactive Nature

By the late 1930s, ecologists had assembled an impressive array of fundamental theories and practical hypotheses to explain nature in a rapidly industrializing world. In this context, connecting theory to practice often challenged these pioneering scientists as they searched for firm evidence. Ecologists suggested that basic components of natural communities followed consistent patterns of distribution and consumption, but they had few direct ways of tracking those movements. The tracking tools they found, ironically, started out as the byproducts of weapons of mass destruction. Near the end of World War II, as wartime priorities shifted to peacetime research, ecologists followed atomic physics into a new era of radiation science. In addition to the new tools at their disposal, ecologists found new questions related to the use and testing of atomic weapons.

The nuclear weapons program known as the Manhattan Project had employed few biologists during the development phase of bomb building. After only a single test in the New Mexico desert, the American military dropped bombs on the cities of Hiroshima and Nagasaki in Japan. The enormous power of these weapons shocked military strategists, political leaders, scientists, and the citizens of the U.S., Japan, and wherever news of the devastation spread. Within days, Japan agreed to unconditional surrender and fears of future nuclear attacks from enemies, both real and imagined, restructured alliances around the world.

Nuclear weapons, once deployed, raised concerns about the effects of radiation on humans and the environment. Like previous bombs, atomic weapons obliterated buildings, destroying property and killing civilians, but they also caused terrible illnesses among survivors. Many observers of the initial effects of the bombs suggested that those killed instantly had been the lucky ones. The unfortunate survivors suffered strange and horrible fates, made worse by uncertainty over the cause of their symptoms. American research teams investigated radiation sickness in Japan. Medical doctors, geneticists, and physiologists who studied the effect of radiation on living tissues in laboratories now had a population of thousands of people in Japan exposed to radiation on an unprecedented scale. The fact that the Manhattan Project had never included a plan to study these effects did not detract from the opportunity to examine them in the aftermath of Hiroshima and Nagasaki.

Concern for survivors suffering from radiation sickness in Japan focused scientists' attention on the human casualties, but some scientists also began to question the broader environmental effects of radiation. These questions developed slowly because opportunities to conduct scientific work in the shadow of further bomb testing were limited by the military need for secrecy. Testing in remote areas, such as small islands in the middle of the Pacific Ocean, also meant that few scientists had ever prepared preliminary studies of plants and animals in those areas. Partly in response to this latter concern, ecologists began to depend upon a kind of study that allowed the broadest kinds of comparisons among living systems. Given the availability of radiation as a tracking tool, ecologists also shifted from the community ecology of Victor Shelford, Frederic Clements, and Charles Elton to a more empirical, systems approach. With evidence for their theories, ecologists and other environmental scientists commanded more respect than their predecessors, but they entered into new arenas of controversy.

In the 1940s and 1950s, concern for human health and the advancement of the nation's economic well being rarely intersected with the need for expanded studies of the environment. While the war stimulated growth in industry and the fields of science that contributed to weapons development, wartime concerns did little to promote ecology and those sciences that had emerged from natural history in the first part of the century. One exception to this trend involved the ecologists who transformed their science to fit the postwar priorities of tracking radioactivity. The ability to transform ecology in this way relied partly on ecologists' development of the *ecosystem* concept, which enabled them to understand nature in terms of a rational system.

Ecosystems

Coined in 1935 by Arthur Tansley, "ecosystem" did not immediately mean the same thing to all ecologists. Tansley's ecosystem concept explicitly included both living and non-living components as essential in defining the larger system. He intended the ecosystem as a unit of study for ecologists. With a defined unit of study, he hoped his colleagues could follow a uniform path toward examining the mechanisms of nature. He believed the ecosystem would simplify the natural world by reducing it to its interacting parts, which could then be studied more like the parts of a machine. It bothered Tansley to see the term ecosystem applied uncritically to communities in a more general way. Many ecologists used "ecosystem" as an alternate term for the community as a superorganism. Tansley had resisted philosophical claims that ecosystem was merely a new name for a living community, which he believed brought ecologists no closer to an understanding of their subject.

Victor Shelford and Frederic Clements, who coauthored a book on *bio-ecology*, wanted to integrate plant and animal studies into one ecological science. They also wanted to create a more precise role for nonliving, or *abiotic, factors* in ecology. Shelford and Clements began to think of abiotic factors as the environmental causes for the structure of a community in a given habitat. Those causes influenced the organisms that would survive and thrive there. For them, ecosystem became another word for community, with a more specific cause-and-effect relationship—the environment was the cause and organisms were the effects. This formulation of the concept, like much of Shelford and Clements's discussion in *Bio-Ecology* (1939), did not win many followers. Although they had directed ecological studies for several decades, their influence had waned by the late 1930s. Ecologists remained uneasy about this loose formulation of the environment. As Tansley had hoped, ecosystem studies soon became a more concrete approach.

The most significant early contribution to ecosystem ecology came from a young researcher whose work appeared in *Ecology* in 1942. Raymond Lindeman presented a view of ecosystems that corresponded closely with Tansley's original intent. Lindeman, who conducted his doctoral research at Cedar Creek Bog in Minnesota, spent a year working with G. Evelyn Hutchinson at Yale. Following his four-year study of Cedar Creek Bog and the year with Hutchinson, Lindeman wrote "The Trophic-Dynamic Aspect of Ecology," an extremely influential article that almost never even got published. Hutchinson lobbied hard for the article with the editors of *Ecology*, claiming its provocative approach pointed in an important new direction for the science. Lindeman's approach fit Tansley's concept of the ecosystem perfectly. He combined the community levels of Elton's food web with a physiological approach that examined food as the movement of energy through a community. In doing so, he rejected any reference to communities as organisms and embraced a view of communities as systems. The organismic views of Clements lacked the rigor of Lindeman's formulation of the ecosystem. His approach also marked a radical departure from Clementsian succession. Lindeman argued that successional change over time resulted not from the growth and maturity of communities, but rather from the changing efficiency of energy transfers between abiotic and biotic factors within the system. The trophic-dynamic approach would enable ecologists to see communities as something other than assemblages of species that vaguely resembled the operations of a hypothetical organism.

Lindeman's trophic-dynamic ecosystem linked the biotic and abiotic factors of the system in a remarkable way. He noted that, previously, ecologists had thought of animals as active players on the stage of plant communities, or of plants as the food source that could influence animal communities. The significance of the abiotic environment for both plants and animals had received little attention. Lindeman believed the environment was the product of all these factors constantly interacting to move nutrients and energy through the system.

Lindeman explicitly included bacterial decomposers as the nutrient-moving part of the system that converted dead tissues from plants and animals back into usable organic compounds. *Decomposers*, which ecologists had formerly all but ignored, enabled nutrients incorporated into living tissue to reenter the system after an organism has died. Without decomposers, nutrients would be trapped, accumulating as dead organic matter. By breaking those tissues down biologically, the nutrients would once again be available to other organisms. This cyclical explanation of nutrients contrasted with Shelford and Clements's cause-and-effect model of abiotic and biotic factors. Lindeman adopted the phrase *trophic-dynamic* to point out the active and ongoing movement of nutrients.

As for energy in an ecosystem, Lindeman extended Elton's community descriptions by examining the depletion of energy at higher levels of the pyramid of numbers. He saw energy as a key to understanding the connection between biotic and abiotic factors. As plants produced usable energy for the community, animals consumed it. Conversion of consumed energy between levels was never perfectly efficient, since much of the energy would be used in foraging and reproduction. At each level, energy loss could not be prevented.

Sadly, by the time his article appeared in 1942, Lindeman had died following a long illness. Hutchinson wrote an addendum to the article, in which he praised his former student for following a line of questions that extended far beyond the immediate problem. The answers Lindeman provided were imperfect, Hutchinson admitted, but they did identify the real problems that ecologists ought to pursue, rather than stopping short by answering only the less significant questions at hand.

After Lindeman's death and the publication of his groundbreaking paper, the concept of the ecosystem received greater attention. It became the basic unit of study for ecologists. Even if it had come to mean more than a Clementsian organism, however, it still carried vague connotations of communities as living organisms. Ecologists could repeat Lindeman's clarifications of energy and nutrient dynamics, but they struggled to add to his substantial claims. In time, new tools and the promise of examining communities as systems merged to propel the ecosystem to the forefront of natural science.

Ecology with a Wartime Focus

In the midst of World War II, with scientific and military efforts to develop nuclear weapons secretly underway, the U.S. government hired fisheries biologists at the University of Washington to investigate radiation effects on major fish species. Lauren Donaldson, who had arrived in Seattle in 1930 to study fisheries at the university and earned both masters and doctoral degrees by 1939, led a group of biologists who studied the effects of radiation on aquatic systems. They functioned under the institutional title of the Applied Fisheries Laboratory created within the university. The publicized goals of the lab were to study the *life cycle* of salmon. The government did not want attention drawn to the radiation work that was being done. Donaldson himself knew nothing of the potential connection between radiation research on salmon and weapons development. The fisheries work was conducted on campus to keep attention away from the site of real concern—the nuclear research at the Hanford facility on the Columbia River in central Washington state, 175 miles southeast of the university.

Donaldson learned about the connection between his fisheries studies and nuclear weapons research in late 1944. By then, his research team had already established a link between salmon egg mortality and elevated radiation levels. When adult salmon encountered even low doses of radiation, a small but significant proportion of their eggs failed to hatch. At higher doses, there were increased malformations among the fish that did hatch from eggs. In the summer of 1945, Donaldson was allowed to move part of the Applied Fisheries Laboratory's research out to Hanford. Interest in how radiation affected aquatic systems expanded, and field studies of radiation became more than an academic curiosity. The Hanford plant had produced radioactivity since at least September 1944, when physicists began testing chain reactions of nuclear materials, the largest created up to that time. These reactions were cooled by water from the Columbia River. Although the reactor was designed to prevent the river from coming into direct contact with the radioactive materials, leaks and accidental releases posed a constant threat.

Sampling around Hanford continued after the war even though much of the government's attention shifted from bomb development to bomb testing. Those biologists with the most experience in sampling for radiation went to the Pacific, where the military planned a series of bomb tests to begin in 1946. In Washington and at other development sites, biologists began to focus on the accumulating radioactive waste. They needed to know how large a threat this waste posed to the health of organisms. The studies at Hanford involved little ecological planning, but biologists began to bring techniques together in ways that stimulated thinking about ecosystems. Richard Foster and J. J. Davis, both fisheries biologists from the University of Washington who had been on the site at the end of the war, followed an *isotope* of phosphorous (P-32) through the food webs of the Columbia River. General Electric ran the Hanford plant and contracted with the university to have Foster and Davis intensify their studies. Knowing that P-32 decays slowly, they found the isotope in increasing concentrations in the tissues of organisms.

Although the amount of radiation leaving the Hanford plant was generally believed to be insignificant, the isotopes lingered in the living system of the Columbia. Small fish, for example, could have P-32 concentrations 150,000 times higher than the surrounding water. Birds were also affected. Young swallows contained 500,000 times the concentrations of P-32 in the water, and ducks and geese laid eggs with 1.5 million times the concentration. Biologists recognized this as an ecological effect of the food chain magnifying low concentrations in the tissues of successive levels of biological consumers. When even the lowest levels of P-32 leaked from the Hanford plant into the Columbia during the process of cooling nuclear reactors with river water, that isotope entered the *food chain* just as any other phosphorous atom would. It became part of the organic molecules making up the tissues of plants and animals, which were then eaten by other animals throughout the food chain. While concentrations of elements like phosphorous remain low in the inorganic environment, the organic tissues of plants and animals retain these elements and accumulate them. Water with traces of radioactivity might remain safe to drink, but eating plants and animals from such environments posed indeterminate risks.

The nuclear operations at Hanford developed into a significant case study for the movement of radioactivity through an ecosystem. Since this was not the original intent of these studies, Foster and Davis had fewer opportunities to explore the environmental effects of radiation. In part, the focus on waste around Hanford resulted from the practical concerns of the contractor operating it, General Electric. The University of Washington became involved at General Electric's request, not to understand the environment but rather to quantify and identify the risks for which the company might ultimately be held liable. At this time, corporations had little to fear from environmental lawsuits or regulations—those realities lay decades in the future—but they did worry about the long-term profitability of highly technical enterprises that might run out of government funding if risks to human health rose too high.

Unlike Hanford, the Oak Ridge National Laboratory in Tennessee was run solely by the government without a corporate contractor. One consequence of that difference was the government's ability to adjust operations to suit a wider range of concerns. During the war, Oak Ridge physicists worked rapidly and regularly improved methods of producing plutonium from uranium reactors. The pace of research and production exceeded any concerns for the safety of the environment. Protecting workers from known hazards was the highest priority but, beyond that, even experienced scientists knew little of the potential effects of radiation. After the war, the pace slowed considerably and safety engineers began to address the looming questions of what might be done with the accumulated waste from five years of radioactive testing and production. As a government owned and operated facility, Oak Ridge administrators could expand scientific operations to include limited studies of the surrounding environment and the potential and actual effects of radioactivity. In 1950, they began to examine the isotopes present in White Oak Lake, a small body of water located on the government site.

The newly formed Atomic Energy Commission directed the work at Oak Ridge. The commission had a branch of health physics and another of biology and medicine, both of which focused primarily on the health of workers and the risks to surrounding human populations. Neither of these branches had hired ecologists to study situations like that around Oak Ridge. Other than the physicists who worked in top-secret labs throughout the war, scientists knew little about the processes of radioactivity and the products they might be studying near those labs. For administrative and scientific reasons, research at Oak Ridge on the site of White Oak Lake proceeded slowly.

In 1954, the Health Physics Division of the Atomic Energy Commission hired Stanley Auerbach to take over the White Oak Lake studies. Auerbach's background as a naturalist and ecologist trained by Victor Shelford made him an ideal person to undertake some of these studies. First, he wanted to find out how varying amounts of radiation might affect well known organisms, such as insects. His studies of relatively simple *dose-response relationships* in arthropods became the first in a long series. Auerbach believed the key to understanding the broader effects of radiation was to begin with small parts of the system rather than study organisms further up the food chain as the Hanford fisheries biologists had done. He focused his research on communities of organisms, as Shelford had always emphasized.

Auerbach began at the insect level in part because of the Atomic Energy Commission's backing, which did not require immediate results that identified human risks. His work resembled that of the slow-paced but consistent Biological Survey. Government scientific administrators actually saw the success of the Manhattan Project as an extremely expensive exception to the rule of steady support for science. Yet, questions of how radioactivity could spread in ecological communities and ultimately affect human health took on a growing urgency. Once initial results confirmed the cycling and accumulation of radioactive products in living systems, the Health Physics Division altered its plan, and Auerbach's patient research on the aquatic ecosystem came to an abrupt standstill. Rather than additional results involving fish, administrators wanted to see more clearly how the risks might reach humans.

Studies of White Oak Lake as an aquatic community ended in 1956 when the Atomic Energy Commission drained the lake. Researchers planted the dried lakebed with corn to see whether crops would take up radiation from the soil. The corn that grew and the insects that fed on it were both contaminated with radioactive isotopes. This result, more like the salmon studies at Hanford, had the potential to bring the danger of radiation home to an increasingly wary public.

The Nuclear Age

When American nuclear weapons designers set out to test their products, they gave little thought to the science that would be needed to analyze the environmental effects of atomic weapons tests. The question of how radiation affected the living world in the aftermath of a nuclear blast, however, quickly required answers. Ecologists had few methods that might help to answer these new and important questions. Analogous studies of radiation seemed particularly inadequate to the broader task of examining the environment. Nuclear physicists, chemists, ecologists, and physiologists combined their knowledge of radiation, ecosystems, and organisms to develop new methods.

A few biologists, mostly physiologists, had studied the movement of radioactive elements within individual organisms. The National Research Council published a volume on these studies in 1936, edited by Benjamin Duggar and entitled *Biological Effects of Radiation*. The book served as a standard reference text for Donaldson and others at the Applied Fisheries Laboratory. These studies used radioactive elemental particles to measure and trace nutrients. These particles came to be known as *radiotracers*. Ecologists began to use radiotracers when waste and fallout from weapons development and testing were released into the environment. While physiologists first used radiation in small, controlled quantities under laboratory conditions, ecologists studying the actual environment had far less control over the systems they studied and the amount or type of radiation involved.

The first bomb test, performed on a remote section of desert in New Mexico known as Alamagordo, involved no ecological surveying or environmental monitoring. The test was a military exercise conducted during wartime to determine whether a nuclear explosion could be initiated and also to find out how much destruction a given amount of fissionable material would yield. The next two nuclear explosions demonstrated the destructiveness of nuclear weapons.

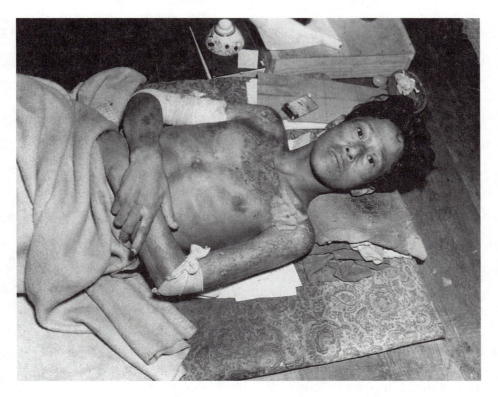

Victim of the atom bomb explosion over Nagasaki, Japan, ca. 1945. (National Archives)

On August 6, 1945, a bomb dropped from a B-29 bomber over the city of Hiroshima, Japan, exploded 1,900 feet above the ground. Its force was estimated to be equivalent to 12,500 tons of TNT. The death toll from such a blast became difficult to estimate. The city held approximately 290,000 civilian residents and 43,000 military personnel. Within a half mile of the explosion, almost everyone was killed. Beyond that radius, nine out of ten people caught outdoors died instantly. By the end of the month, close to 100,000 people had died as a direct result of the blast or from the radiation poisoning that followed. Complications from exposure may have killed 200,000 people by the end of 1950. Of the 76,000 buildings standing in Hiroshima on August 6, 48,000 were totally destroyed, and all but 6,000 were damaged.

Three days later over Nagasaki, another B-29 dropped a bomb that exploded with an estimated force of 22,000 tons of TNT. The steep slopes around the city confined this larger blast, but again, tens of thousands of people died. By the end of December, estimates put the death toll at 70,000. Over the next five years, another 70,000 died of complications from the blast and its radiation.

The two bombs dropped on Japan ended the war. They also provided a gruesome opportunity to observe the results of nuclear explosions and radioactivity in a populated area. These effects received no scientific study beyond the human health concerns. Ending the war by these means, however, created an ongoing military interest in atomic weapons, especially since the Soviet Union also possessed the means of producing these bombs. Shifting alliances after the war heightened tension between the U.S. and the U.S.S.R., and the threat of a nuclear attack eventually became a weapon of fear that many strategists believed would be the only means of preventing just such an attack. The period came to be known as the *Cold War.* As a result, the U.S. nuclear program expanded after Hiroshima and Nagasaki.

Testing Sites

The Atomic Energy Commission and the U.S. military began planning additional tests of atomic weapons after the end of World War II. Lauren Donaldson left Washington state for the Pacific when his expertise with aquatic systems and radiation became useful for monitoring the effects of bomb tests. Planning for tests demanded careful consideration of where the risk of nuclear fallout could be minimized. In essence, this meant simply reducing the risk to human life. Looking across the vast Pacific Ocean, the remote Marshall Islands became leading candidates. Most of the islands consisted of little more than a coral reef that emerged above sea level only during the lowest tides. These islands, called atolls, had a circular or oval shape surrounding a central lagoon, outlining the tops of volcanic cones that rose 15,000 feet or more from the ocean floor. In most cases, the original volcano had eroded away, leaving only narrow bands of living coral, filled with sand and limestone. Sections of the atolls could support scrubby vegetation and even palm trees, but very few people. The islands were thousands of miles from Hawaii, but not too close to more densely populated islands near Asia. Bikini Atoll, the location of the first Pacific atomic tests, was home to only 162 human inhabitants. The U.S. government moved these people to another island, with plans to return them when the testing was complete.

The first tests of atomic bombs after World War II took place in July 1946 on Bikini Atoll. The military assigned them codenames Able and Baker. Test Able involved the detonation of a bomb 518 feet in the air above a fleet of retired naval vessels. The blast sank five ships, the first within one minute, and the last within 24 hours. By the time the fifth ship sank, scientists and naval troops began recovering scientific instruments from the remaining vessels. They also began examining animals that had been left aboard the ships to test the effects of the blast and radiation. Test Able left levels of radiation on the ships and in the surround-

View of "Baker" atomic bomb explosion at Bikini Atoll on July 25, 1946, the last of three American tests. The blast sent up a column of water 5000 feet high and 2000 feet wide at the base. (Corbis)

ing lagoon that observers considered surprisingly low given the intensity of the explosion. Test Baker, which the military detonated underwater, ninety feet below the surface of the lagoon, left higher radiation levels. The levels of that blast exceeded the expectations of nuclear scientists. The underwater blast also sank five ships, along with two submarines, and it also created a radioactive sludge on the lagoon floor that left long-term consequences for the biotic community far beyond the atoll itself. No one had foreseen the complexity of problems this radioactivity could cause, so the lagoon became a laboratory for marine study of these effects.

Biologists had begun monitoring fish species in June, prior to the first test at Bikini. They collected 2,000 fish from the atoll before June 29. Between the tests, they collected nearly 2,000 more. After Test Baker, officials considered radiation levels to be too high to safely collect more fish until a week had passed. They hoped the radiation would be confined to the lagoon. Scientists collected another 1,400 fish, ending their sampling in mid-August, by which time radioactivity had entered the atoll's food web at all levels. The thousands of species of protozoa at the base of the food web contained elevated levels of radiation, as did as many as 1,000 species of fish. Some of these fish were herbivores, eating

Bikini Atoll, the Floating Phantom of the Pacific: Sharks swim around the remains of the USS Saratoga *which also housed "Curtiss Helldiver" dive bombers stationed on the flight-deck at the moment of the bomb's impact. (Corbis)*

only marine plants, but most were carnivores that ate other fish. Among the carnivorous species, shark and grouper could swim great distances, broadening the impact of the tests as they carried traces of radiation in the tissues of their bodies. The initial studies of the biological uptake of radiation, though valuable, were quickly seen as crude and inadequate.

A report written by Arthur Welander from Bikini Atoll described the biological findings of the first samples taken after Test Baker. Welander noted that none of the biological specimens taken within the lagoon remained entirely free of radioactive contamination. He found algae to have the highest levels of radioactivity, which explained why fish and other animals feeding on the algae also showed high radiation levels. Reemphasizing the unexpectedly high levels of fission production throughout the lagoon, and the biological spread of radiation beyond, he warned against using fish, mollusks, or other animals from within 100 miles of Bikini for food.

With Tests Able and Baker complete, the Bikini Atoll offered limited additional usefulness to the military. The preliminary sampling by biologists confirmed the spread of radioactivity from the two tests and suggested that underwater detonations like Baker would prove unnecessarily risky in concentrating fission products. The sampling, for the purposes of the military, had shown that ongoing testing would spread radiation in the environment. Since the military considered additional tests essential, they now had at least gained some insight that might help limit contamination. Officials scrapped plans for further underwater tests. The long-term consequences of radiation remained a topic of speculation, but the surveys conducted immediately after testing also showed the resilience of organisms. Remarkably, Bikini Atoll appeared unchanged in its biological essentials. Apart from the presence of radioactivity, its food webs remained intact.

When the Atomic Energy Commission took control of the nuclear program after the war, the shift from development to testing was on. The Manhattan Project gave way to testing new and more powerful bombs, searching for better ways of converting nuclear fission into destructive power that could deter enemies from even considering an attack on the United States. Lauren Donaldson's Radiobiology group became part of the Atomic Energy Commission in 1947. He returned to Bikini to collect additional samples, but plans for further atomic tests elsewhere had already been made. The next tests would take place on Eniwetok Atoll, a larger group of islands where the military could build adequate landing strips and construct a permanent base. Having learned that above ground tests caused less contamination of marine life and sediments, the Atomic Energy Commission planned to detonate three bombs atop two-hundred-foot tall towers. The tests began on April 15, 1948, with a second blast on May 1, and a third on May 15. President Truman announced the Eniwetok tests to the public two days after the last detonation.

Because of his experience on Bikini, Donaldson observed the three tests in 1948, but conducted no extensive survey of marine life. He made arrangements to collect a few samples from the edge of the fallout zone with a small team. The team retrieved tissue samples from 118 specimens of various species and transported them back to the lab in Seattle. Donaldson found high levels of radiation, but they had little information with which they could compare these levels. When the Atomic Energy Commission established a more extensive sampling program, the Radiobiology Group conducted additional surveys for less than a month each in the summers of 1948 and 1949.

Bikini and Eniwetok atolls became field stations for radioactive testing of biological samples, but biologists had little idea what they were looking for or whether their findings were significant. In addition to the fact that no biologists had studied these areas of the Pacific in detail before testing began, specifics about the bombs themselves remained classified even to the biologists who worked on the project. It became increasingly apparent that whatever quantitative data researchers collected would have little meaning against a blank background of unknown baselines. No one knew in any detail what the ecological food webs of these atolls looked like before testing. After the sampling in the summer of 1949, biologists could only conclude that radiation levels continued to decline, but also that radiation persisted. It would not reach zero any time soon, as least not in the tissues they sampled for comparison. The studies concluded in uncertainty, and with the threat of war in Korea looming, priorities shifted rapidly in the Pacific during the next year. The military put bomb testing on hiatus, and biological monitoring on the atolls all but ceased.

The Atomic Energy Commission continued to expand its programs in the early 1950s as priorities in Korea became more clear. The expansion of atomic programs included ongoing funds for the Applied Fisheries Laboratory, although biological testing remained a small piece of the larger plan to understand atomic power.

Ecological Studies in the Pacific

In the midst of massive construction on Eniwetok Atoll, the U.S. government established a small marine biological laboratory and began operations there in 1953. By then, two ecologists, brothers Howard and Eugene Odum, had implemented Donaldson's plans to follow the movement of radioactive isotopes on and around the atoll. Howard Odum had earned undergraduate degrees in chemistry and zoology before completing his doctoral research under G. Evelyn Hutchinson at Yale. He studied the levels of a naturally occurring isotope of strontium in ocean water, drawing some bold conclusions about the constancy of global conditions over millions of years. He had also served as an Air Force meteorologist during World War II. Eugene Odum studied ecology at the University of Illinois with Victor Shelford. With more training and experience as a field ecologist than his brother, Eugene brought the perspective of a naturalist to Eniwetok. Together, they had already authored the first important textbook of ecology. Having almost no experience with marine organisms or radiation, however, they felt ill-equipped to study coral reefs. Their approach, based on their combined expertise in biogeochemical cycles and food web analysis, nevertheless yielded an immensely successful scientific paper. The research methods developed by the Odum brothers, more than their familiarity with the particular ecosystem, provided a model for ecosystem studies thereafter.

In 1955, the Odums published a paper that examined a reef section of the atoll where nuclear explosions had not affected marine life to any visible degree. For the first time on a nuclear test site, this study was not primarily to determine the impact of an atomic blast. Instead, the Odums realized their unique opportunity to examine the effects of radiation on whole populations and an entire ecosystem in the field. They described the flow of energy and the cycling of nutrients on the reef without further reference to the massive atomic tests conducted by the military in recent years. For example, they found that coral species play a complex and dynamic role in capturing energy for use in the community by serving as a host to diverse algae species. Coral also took in nutrients directly and reused those nutrients internally. These processes meant that radiation could enter the many members of the marine community by more avenues than originally recognized by ecologists, and that radiation could remain within the ecosystem much longer than anticipated.

In November 1952 on Eniwetok, Test Mike included detonation of the first thermonuclear bomb. This new type of nuclear weapon used the uranium and plutonium fission reactions of earlier bombs to ignite a fusion reaction between isotopes of hydrogen. The result was a much more massive explosion. In March 1954, Test Bravo coincided with the unexpected presence of a Japanese fishing boat in the fallout zone. Thus the Bravo blast tragically provided further evidence of radiation dangers. The crew members became severely ill during the two week voyage home, and one later died of complications from the fallout exposure. As a further consequence of that incident, Japanese citizens felt a new sense of fear over radiation and American atomic bombs as concerns increased about the safety of fish for human consumption.

The Odums steered clear of these issues, however, and focused on the movement of energy and nutrients in the affected ecosystems. The sampling done after the explosions had followed the protocols of earlier monitoring. Investigators recorded radiation levels in tissues of various organisms and noted the damage to vegetation at varying distances from the test site. They calculated the efficiency rates at which organisms could use the nutrients provided by various food sources. Where Charles Elton had drawn pyramids of numbers of organisms at different levels of the food chain, Howard and Eugene Odum developed more detailed methods of measuring and estimating the biomass of different species. *Biomass* included the weight of dried organic matter from a particular habitat. The Odums took separate measurements for plants (*producers*) and other organisms (*first- and second-level consumers*). They could use these measurements to compare the efficiency of the Eniwetok ecosystem to other familiar ecosystems. These examples became textbook cases of how ecologists could use radiation as a tool to identify connections within diverse ecological settings.

The movement of radiation through an ecosystem provided an extremely useful method for ecologists to study nutrient cycles and the efficiency of natural systems. In turn, ecologists suggested a means for studying the distribution of radioactive fallout and its potential effects on humans. Radiation measurements on Rongelap Atoll, a ring of islands east of the test sites on Eniwetok, revealed the movement of radioisotopes across the open ocean. Radiation went

Eugene (1913–2002) and Howard T. (1924–2002) Odum

As a teenager, Eugene Odum began writing a weekly column on birds for the *Chapel Hill Weekly*. His writing career blossomed not only because he worked hard at communicating but especially because he had profound ideas to share about the natural world. Many of the concepts proposed by ecologists in the 1920s and 1930s found a much wider audience after Eugene revised and integrated them into larger frameworks, which he ultimately published as the first general textbook of ecology in 1953. Part of the inspiration for Eugene's *Principles of Ecology* came out of a frustration that his colleagues at the University of Georgia, where he joined the faculty in 1940, saw ecology as an unimportant and narrow sideline within biology. He helped to make that sideline one of the most significant and broadly interdisciplinary fields of modern science.

Eugene P. Odum became a pioneer of ecosystem ecology and brought unprecedented government funding to ecological research in the 1950s and 1960s.(Getty Images)

Raised in Chapel Hill, North Carolina, Eugene earned bachelors and masters degrees in zoology from the University of North Carolina, and a doctorate from the University of Illinois. His father, Howard Washington Odum, was a famous sociologist, and his mother was an urban planner. His younger brother, Howard T. Odum, also became an ecologist, and together they coauthored the first edition of *Principles of Ecology*. The brothers shared the Crafoord Prize of the Royal Swedish Academy in 1987 for their work in pioneering ecosystem ecology. The prize is the highest honor an ecologist can receive, equivalent to a Nobel Prize, which is not given in ecology.

As a boy, Howard Thomas Odum taught himself the basic principles of electricity. He learned about circuits and resistors, becoming keenly aware of how electrons move through electric systems. The younger Odum's interest in electricity, combined with his brother's knowledge of birds, made them a scientific team. On many days after school, Howard also worked for a marine zoologist at the university, so it was no surprise when he chose zoology and chemistry as a double-major in college.

After college, Howard began his first studies of biogeochemistry as a graduate student. He studied the cycling of a single element, strontium, in the Earth's oceans. He

from water to fish, from soil to plants, to other organisms, and then on to humans. Edward Held published his findings from Rongelap as part of the Proceedings of the First National Symposium on Radioecology in 1962. Examinations of the movement of fallout closely mirrored studies of ecosystems. Improvements in ecologists' measurements and methods allowed closer tracking of the potential radiation effects on humans.

determined that the chemical composition of the oceans had not changed in the past forty million years. He could not have known then that strontium would become an element of major concern in nuclear fallout, but his investigations and the methodology he established helped him become one of the foremost experts in studying the effects of fallout in the 1950s.

Like his brother, Howard made significant contributions to the development of ecosystem ecology. His particular interest in electricity led him to examine the movement of energy through ecosystems as analogous to the movement of electrons through circuits. He reconfigured the energy levels of the pyramid of numbers into boxes on ecosystem diagrams similar to resistors on circuit diagrams. Howard drew inspiration for these new representations of nature from the carbon cycle charts plotted by his graduate professor G. Evelyn Hutchinson. By bringing together ideas from these different approaches, Howard helped to pioneer ecosystem ecology.

Among Eugene Odum's many contributions to the science of ecology, his role in founding the Institute of Ecology at the University of Georgia remains a hallmark. Today, the often imitated institute employs 150 researchers and assistants with a budget of over $12 million. When Eugene served as its first director, it consisted of a few graduate students and a $10,000 government grant. He also helped establish the Savannah River Site, a 300-square-mile ecological laboratory of both pristine and heavily contaminated areas. The Marine Science Institute on Sapelo Island on the Atlantic Coast also owes its existence to Eugene Odum's founding efforts.

Howard Odum spent his career at the University of Florida. His first major book, *Environment, Power, and Society* was published in 1971. He increasingly advocated a restructuring of society based on principles of science and technology. His last book, *A Prosperous Way Down: Principles and Policies* (2001), continued his call for rational restructuring of society and improved energy policy planning. He died in 2002, one month after his brother.

For his contributions to ecology as a writer, a teacher, an institution builder, and an innovator, Eugene Odum is remembered as the "father of modern ecology." When he died in 2002, obituaries claimed that he had made ecosystem a household word, and that references to the "circle of life" in Disney movies could be traced to his insights. Such claims, exaggerated in the reverent convention of honoring the recently deceased, do point to the important lifelong contributions of this tireless ecologist.

Global Awareness

Ecologists were not alone in recognizing the enormous impact humans could now exert on the planet. At a conference in 1955, scientists, urban planners, and governmental consultants met in Princeton, New Jersey, to discuss "Man's Role in Changing the Face of the Earth." Their primary aim was to make public the

Aldo Leopold (1887–1948)

Often referred to as the "father of game management," Aldo Leopold came to wildlife science after starting his career in forestry. He wrote *Game Management* (1933), the

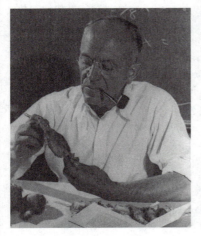

first textbook in the field, after synthesizing recent ideas in animal ecology with what he had learned during a three-year survey of game laws and wildlife populations in the Midwestern states. The result provided a practical guide to game management that went beyond wildlife protection efforts from the turn of the century.

Leopold grew up in the Midwest and developed a love for hunting. His home in Burlington, Iowa, sat near the Mississippi River, and in the wooded lands along the river, he hiked and hunted with his father and brothers. His grandfather, an avid gardener, taught Aldo to be an enthusiastic naturalist. He decided to become a

The influential writings of 20th-century environmentalist Aldo Leopold (1887–1948) changed the way that many Americans viewed nature. An employee of the U.S. Forest Service, Leopold was instrumental in the creation of Gila National Forest, the nation's first protected wilderness area. (Library of Congress)

forester, and in preparation for entry into the Yale Forest School, he left home to attend a boarding school in New Jersey. In the years away from his family as a teenager, Leopold refined his writing style by sending letters home and allowing his mother to offer a critique of each one.

By the time he finished his training at Yale in 1909, Leopold hoped to join the U.S. Forest Ser-

contributions of their diverse fields. They did not intend to catalog the ways in which human activities had changed the globe, for better or for worse, although they did assemble commentaries from leading scientists on that subject. They concluded that a multidisciplinary approach that applied science to urban, social, and natural problems could combine technology and conservation. Scientists in many fields could work toward integrated solutions, in contrast to the predominately technological methods of the previous fifty years.

Among the seventy-five contributors to the published volume that came out of the symposium, five came from departments of anthropology, eighteen from geography, five from the humanities, six from botany, three from ecology, and two from zoology. The remainder ranged from land planners to research directors. The organizers of the symposium included geographer Carl O. Sauer, zoologist Marston Bates, and land planner Lewis Mumford. They not only hoped to combine the insights of a variety of disciplines, but they also selected participants for the individual contributions each might make.

vice in one of its most remote regions. His wish was granted and he was assigned to the Arizona Territory. He worked his way through several projects, making errors in calculations and struggling to supervise his team. They were charged with surveying and estimating the amount of timber on the public lands, but Leopold's inexperience plagued him. He quickly realized that his strengths in writing and organizing ideas did not necessarily help him as a forest supervisor. The Forest Service eventually placed him in a desk job after a serious illness nearly killed him.

In 1915, Leopold became interested in the way the Forest Service administered the game animals on its lands. He believed that the government should manage its birds and large mammals according to scientific principles, just as it did with trees. His job did not allow him to spend much time on this idea. Writing became his way of contributing to the science of game management, and he increased his involvement after moving from the Southwest to Madison, Wisconsin, where he eventually left the Forest Service and began a new position at the state university. Around the time *Game Management* was published, Leopold became the first professor of game management in the country and, in 1939, the university established a Department of Wildlife Management.

Leopold spent the 1940s teaching, battling politicians to keep the state's game department out of the hands of special interests, and restoring the land around a small shack on a piece of property along the Wisconsin River near Baraboo. As part of each of these activities, he increased his daily writing routine. He wrote research articles as a professor, he wrote policy guidelines as a game manager, and he wrote a diary of his efforts to grow grasses and trees on the worn out land he had purchased. The best known of his writings were published as *A Sand County Almanac* in 1949, the year after Leopold died fighting a grass fire near the shack.

Paul Sears, an ecologist, expressed his belief that the natural and social sciences had a great deal to contribute to the betterment of society and the conservation of the earth's resources. Sears began by asserting that people in the twentieth century saw their place in nature as different from the place of other species. Many scientists, driven by the promise of modern technology, remained hopeful that humans would face few limitations in the future, despite the recent lessons offered by nuclear fallout. Such optimists predicted only positive results from scientific knowledge, but Sears wrote as an ecologist to remind them of the existence of limiting factors in nature.

Marston Bates criticized ecologists for not considering the role of human activities more consistently in their work. For example, a species that existed as a dominant grazing animal throughout many areas of the world, the domestic goat, did not appear in any of several ecology textbooks he examined. Yet, Bates argued that goats played a major role in the ecological relationships being studied in those areas. This narrow approach represented a failure to recognize the range of human activities and their effects on ecological systems. The call for ecologists to recognize human factors in their studies challenged those scientists to acknowledge the limits of their profession. At the same time, as global changes became more clearly coupled with nuclear tests, large-scale agriculture, pollution, and resource extraction, scientists saw enormous opportunities for ecology to develop into an important source of expertise for facing the future.

Leopold's Philosophy

Even before the dramatic potential of nuclear power for altering natural and cultural landscapes was realized, Aldo Leopold had begun reflecting on how humans viewed nature differently than previous generations. Amid wartime preoccupations with global events, enemies in Europe and Asia made Americans' relationship with nature seem more benign and less urgent than at any time since at least the mid-nineteenth century. As the war drew to a close, however, and scientists looked to the future, they faced a wider range of possible subjects for study than ever before. In addition to funding increases and studies of ecosystems, a few individuals posed remarkable questions about what their work might mean to future generations. Leopold, who began as a forester before turning to game management, emerged as a voice for considering the future of human relationships with nature. His reputation as the father of game management shifted to that of the prophet of responsible land use.

Leopold, like many ecologists, had grown more interested in the structure of ecological communities and the makeup of ecosystems. He saw how humans disrupted natural communities, and those disruptions troubled him as a scientist, as a hunter, and as a philosopher. He combined the impulses of these three components of his personality in a stream of essays during the 1940s. Although not trained as an ecologist himself, he followed ecological publications closely and drew a range of lessons from their conclusions. He understood that human actions could destroy the habitat for particularly vulnerable species and those actions became magnified by wartime resource extraction and postwar development.

Leopold examined the contrast between modern development and rustic renovation when he bought a small piece of property in 1935. He called the place the "Shack" after the dilapidated building left barely standing by the previous owners and began to spend his weekends there, away from the University of Wisconsin. With the help of his family, Leopold hoped to restore the vitality of the natural community that once thrived there. He had no intention of farming the land. In fact, he blamed farming for the worn out condition of the sandy soil. He wanted to undo the damage of generations of farming. At the same time, he documented the effort that went into the project. The fact that Leopold died in 1948 fighting a grass fire at the Shack made him a martyr to his quest.

Among the essays Leopold wrote in the 1940s were a series examining the changing of the seasons at the Shack. He began to refer to those essays as an almanac. Some represented little more than journal entries. When he died, the essays were published as *A Sand County Almanac*, a testament to the soil of that area of south-central Wisconsin. Along with the record of seasonal changes, a number of essays on a wider range of topics were included in the book. The metaphors and imagery of these became legendary.

In an essay entitled "Thinking Like a Mountain," Leopold described the damage done by herds of deer to mountains unprotected by wolves. He had taken the case of the Kaibab deer and liberally interpreted its meaning. The lesson for a new generation of conservationists, Leopold believed, was that people must learn to think like mountains, to take a longer view of the changes wrought by human disturbance of ecosystems, even to welcome the actions of predators. When humans removed components of ecosystems, such as predators, they could hardly be sure of the outcome. Deer, protected from their enemies, would overpopulate their range, destroy its vegetation, and starve. The lesson that humans should think on time scales larger than one lifetime, by then introduced into a few ecology textbooks, became ubiquitous after Leopold tied it to an appealing and logical metaphor. Ecologists ever since have struggled to disentangle the common sense metaphor from the more complex ecological reality.

In another of his essays, Leopold expressed a notion called the *land ethic*. This extension of ethics had its basis, Leopold believed, simultaneously in biblical prohibitions against the exploitation of land, in limits set by the Darwinian struggle for existence, and in concepts of community ecology. Leopold's land ethic included soil, water, plants, and animals in the same way that ecosystems encompassed abiotic and biotic factors. Scientists' growing awareness of the complexity of relations among these factors made him less confident that they would ever know in detail what made ecosystems function. Leopold believed farmers and landowners alike needed to develop an "ecological conscience" to overcome the self-interested practices that yielded economic gain at the expense of ecological destruction. Without such a conscience, economic arguments would suffice to protect only a small percentage of the species that actually kept communities functioning. In Wisconsin, Leopold estimated that only about 5 percent of the 22,000 species of plants and animals would be put to economic use as lumber or agricultural products. Because economic incentives were paramount, and because governments were loathe to interfere with those incentives on behalf of the land, Leopold made his appeal to the conscience of the private land owner. Leopold summed up the land ethic as follows: "A thing is right when it tends to preserve the integrity, stability, and beauty of the biotic community. It is wrong when it tends otherwise" (*Sand County Almanac*, 224–225). This principle of land management became a fundamental concept of environmentalism for decades afterward. He hoped the conservation movement of his day would be the embryo of an ethic that would include proper land use.

Placed just before "The Land Ethic" in *A Sand County Almanac* was an essay on wilderness, with which Leopold arguably intended to conclude the book. In that essay, he reflected on the uses of wilderness for recreation, for wildlife, and for science. He described how wilderness lands had become marginal. Given the flawed imperatives of economics outlined in the land ethic, the promise that wilderness might hold for future generations could not be realized. In particular, Leopold argued that wilderness, even small remnants, ought to be preserved as study areas—even laboratories—for comparison. He suggested that each biotic province requires an example of its own wilderness for comparative studies. Ecologists would not learn how an ecosystem in Montana functions by studying the Amazon, so each place would need its own preserves. Leopold offered examples of what ecologists had recently learned about plant growth in agricultural areas, where root systems drew moisture and other nutrients from only one level of the soil. In native prairies, grasses and wildflowers distributed their roots to multiple levels of the soil, neither overdrawing from one nor neglecting others. Without access to a native prairie, ecologists could not establish such a study. He believed that wilderness could thus assist, through science, practices in agriculture and forestry. "In many cases," Leopold added, "we literally do not know how good a performance to expect of healthy land unless we have a wild area for comparison with sick ones. . . . In short all available wild areas, large or small, are likely to have their value as norms for land science. Recreation is not their only, or even their principal, utility" (*Sand County Almanac*, 197–198).

Leopold's work, based largely on the ecological science of his day, expanded the relevance of that science. His audience was not large in the late 1940s, and he might have continued to refine his ideas for another decade, had he lived longer. When *A Sand County Almanac* appeared in 1949, few ordinary citizens were familiar with the ecological work of Charles Elton or Victor Shelford, and few ecologists felt they could be bothered with the somewhat preachy philosophizing of Leopold. By the 1960s, however, the audience he had imagined began to materialize. In particular, the spread of radiation and the effects of toxic chemicals illustrated the destructive and far-reaching effects of human activity. Leopold's appeals to conscience and ethics became meaningful responses to the fear people began to feel. Many observers have since remarked that his ideas were ahead of their time. In fact, the mindset that he had imagined of his readers developed out of the synthesis of those ecological ideas in subsequent decades. Only then did his work earn the appreciation of a wider audience.

Conclusion

The convergence of nuclear testing with environmental science transformed ecology from a professionalized specialty of natural history into a twentieth-century systems science. Ecologists found opportunities for funding that would have seemed impossible a few decades earlier. By the end of the 1950s, a few programs continued under the direct administration of the Atomic Energy Commission, particularly those at the national laboratories, such as Oak Ridge. Other programs emerged with less oversight from the government and military. At Savannah River in Georgia, Eugene Odum turned a three-year federal grant for $10,000 per year into a long-term research program. Rather than focus on the potential hazards of a nearby nuclear plant, Odum hired graduate students to conduct ecological surveys using radioisotopes to trace nutrients. They remained committed to ecosystem studies instead of getting caught up in the race to win funding from the Atomic Energy Commission. The government might have preferred that Odum and his students focus on radiation studies but, by 1960, he had secured enough funding to build an onsite lab. The Savannah River Ecology Laboratory, officially a component of the University of Georgia, became a premier site of ecosystem studies for years afterward.

Different questions had motivated Lauren Donaldson, who wanted to improve fisheries in the Pacific Northwest, but he pursued a similar tactic by conducting studies with radiation. He created a long-term study at Fern Lake, a small body of water that supported almost no life as a result of heavy rainfall on volcanic rock that retained few nutrients. Located southwest of Seattle, across Puget Sound, Fern Lake provided what Donaldson believed would be an ideal natural laboratory not far from the University of Washington. From 1957 until well into the 1970s, Donaldson and his students examined every aspect of the lake's ecosystem. These researchers not only hoped to understand the food web of this place, they wanted to learn how to make it more productive. Fern Lake was in many ways typical of northwestern lakes, and the inability of many of these lakes to support salmon populations provided an intellectual as well as an economic challenge to ecologists. Donaldson suggested that by improving mineral and nutrient distribution in Fern Lake, the means of producing fish in many such lakes could be discovered. What he found instead was that the natural complexity of aquatic ecosystems ultimately frustrated these practical aims. In the process, however, researchers at Fern Lake refined techniques for tracing radioisotopes and contributed significantly to aquatic ecology.

These examples demonstrate how the commitment to a basic understanding of nature drove scientists during this period to develop new methods and concepts with the expanding set of tools available. While the war and the political climate of the postwar world provided funding for these new studies, ecologists rarely chose to focus narrowly on the public's growing concern over human health. The Atomic Energy Commission initially supplied funding for those narrower questions, but through the support of their academic communities and the ingenuity of establishing small-scale laboratories with shoestring budgets, many ecologists managed to continue with a focus on broader ecosystem subjects.

Ecologists combined interests in the natural world with concerns about how changes might affect human well-being. In fact, as the hazards of radiation became more evident, scientists worked collaboratively to make those dangers known and to investigate the persistence and spread of the products of nuclear testing. Radiation quickly became a symbol of the danger that human activities created for other species as well as for every man, woman, and child. Identifying a wider range of environmental risks soon occupied a large proportion of scientists' energies. As suggested at the interdisciplinary conference on how human activities had changed the face of the Earth, ecologists needed to direct increasing attention toward humans in their studies. Individually, Aldo Leopold probably had as much influence on ecology as anyone in the decades that followed. His eloquence and candid concern on behalf of the environment inspired a generation or more of scientists and an even broader cross-section of the public that came to think of themselves as "environmentalists."

Bibliographic Essay

The most comprehensive history of nuclear testing, which includes some discussion of biological and ecological sampling and assessment, is Neal O. Hines, *Proving Ground: An Account of the Radiobiological Studies in the Pacific, 1946–1961* (Seattle: University of Washington, 1962). On studies of the Atomic Bomb Casualty Commission in Japan after the war and the work of geneticists to interpret the meaning of those studies, see John Beatty, "Genetics in the Atomic Age: The Atomic Bomb Casualty Commission, 1947–1956" in Keith R. Benson, Jane Maienschein, and Ronald Rainger, eds., *The Expansion of American Biology* (New Brunswick: Rutgers University, 1991), 284–324. The most complete examination of Lauren Donaldson's work at Fern Lake is Matthew W. Klingle's "Plying Atomic Waters: Lauren Donaldson and the 'Fern Lake Concept' of

Fisheries Management" *Journal of the History of Biology*, 31 (1998), 1–32. For more on the controversial and complex events surrounding radiation safety during the development of the first atomic bombs, see Barton C. Hacker, *The Dragon's Tail: Radiation Safety in the Manhattan Project, 1942–1946* (Berkeley: University of California, 1987). Among the best recent histories of what later came to be seen as environmental activism in the 1950s is Mark W. T. Harvey's *A Symbol of Wilderness: Echo Park and the American Conservation Movement* (Seattle: University of Washington, 2000). In this book, Harvey distinguishes the popular activism based on the aesthetics of wilderness of the 1950s from that based more substantially in the environmental science of the 1960s and later.

Eugene Odum's work includes the pioneering textbook, *Fundamentals of Ecology* (Philadelphia: Saunders, 1953), which was republished in 1959 and 1971. A coauthor on the first edition of that textbook, Harold T. Odum wrote several important books, including *Environment, Power, and Society* (New York: Wiley-Interscience, 1970). Lauren Donaldson's work has been collected in *The Fern Lake studies [by] Lauren R. Donaldson . . . [et al.]*, ed. by Marion L. Chase (Seattle: University of Washington, 1971).

An excellent and readable account of Aldo Leopold's life is Marybeth Lorbiecki, *Aldo Leopold: A Fierce Green Fire* (Helena, Montana: Falcon, 1996). The most complete biography of Leopold is Curt Meine's *Aldo Leopold: His Life and Work* (Madison: University of Wisconsin, 1988). Additional and more accurate insights into the development of Leopold's thinking are provided by Susan L. Flader in *Thinking Like a Mountain: Aldo Leopold and the Evolution of an Ecological Attitude toward Deer, Wolves, and Forests*, 2nd ed. (Madison: University of Wisconsin, 1994).

6

Environmental Science and the Challenge of Environmentalism

In the 1960s, environmental science began a dramatic interplay between political and social concerns. That interplay arose largely from public announcements by scientists that human activities threatened the natural world. The concern for the environment that developed outside of environmental science into a political and social movement, *environmentalism*, focused on identifying problems and proposing solutions. Environmentalists and environmental scientists represented separate but overlapping communities. Environmentalists needed the empirical work of environmental science to advance their political and social agendas. While some environmental scientists participated actively in pushing these agendas, others saw environmentalism as a threat to the objectivity of environmental science. The mutual dependence of these communities created an unresolved tension that lasted over a generation, and still lingers today.

Environmentalism changed the way environmental scientists approached studies in ecology, geology, and meteorology, as well as chemistry, biology, and geography. The acknowledged intersections of these sciences were often cited as the potential solution to polluted air and water or disappearing species. The public sometimes assumed, however, that gaps between the sciences represented areas of ignorance that could lead to environmental catastrophe. Scientists regularly confronted the aspects of their work that attracted the attention of an awakened public, but at times they attempted to avoid environmental controversy. When they took on controversial questions, the broader public was often unprepared to deal with nature's complexity and they required significant education about scientific methods and results. Those who succeeded in reaching a wide audience, like Rachel Carson, often faced critics on one side who challenged her scientific conclusions and detractors on the other side who dismissed her as a mere popularizer. The paths that scientists found through this era of heightened interest in their work illustrate the sometimes difficult relationship between science and society.

Environmentalism did not arrive on the scene as the result of a single, dramatic event. Rather, a number of sparks ignited a spreading fuel of awareness. Rachel Carson's *Silent Spring* (1962) created an unprecedented public and political response that has earned it the reputation of initiating concern over pesticide use and toxins in the environment. Oil spills and construction proposals that threatened human health and wildlife in the mid-1960s also inspired activism on behalf of the environment. Important legislation passed in the U.S. Congress included Clean Air Acts (1963, 1967, and 1970), Clean Water Acts (1960, 1965, and 1972), and Acts for the protection of endangered species (1964, 1968, and 1974). Citizen action around the country on Earth Day in April 1970 also marked significant origins of the environmental movement.

DDT as the First Fear

Dichloro-diphenyl-trichloro-ethane, which was first synthesized in Switzerland in 1874, gained popularity as an insect killer when its effectiveness for that purpose was demonstrated in 1939. By the end of World War II, it was known as *DDT* and became widely available. It seemed harmless to wildlife and to humans, and consequently became a promising weapon against insects. The U.S. Department of Agriculture recommended its use to American farmers as a safe way to increase crop yields that would otherwise be eaten by pests.

Experiments on the effectiveness of DDT caught the attention of a biologist in the Fish and Wildlife Service. Her name, Rachel Carson, eventually became synonymous with criticism of widespread use of DDT. In 1945, Carson proposed an article on the chemical, focusing on research programs initiated by the Department of Agriculture, the Public Health Service, the Food and Drug Administration, and the Fish and Wildlife Service. The first three agencies ran a joint program known as the Agricultural Research Service in Maryland, while the Fish and Wildlife Service had its own facility nearby. Carson worked primarily as a writer and editor for the government, but she proposed the article for *Reader's Digest*. Since the Agricultural Research Service had much more funding and would undoubtedly present its findings on the chemical as being beneficial to agriculture, Carson believed the story of the effects of DDT on other animals might prove an interesting counterpoint. The editors did not share her interest, so the article was never written. Over the course of the next decade, however, she continued to monitor the research. As she had suspected, the Fish and Wildlife Service research soon suggested that DDT was not harmless.

Carson did not conduct research of her own on synthetic pesticides. During the 1950s, she continued to pursue writing projects beyond her position as

A ranch hand sprays cattle with a DDT solution in a holding corral in 1947. Rachel Carson wrote of the dangers of the pesticide in her groundbreaking book, Silent Spring *(1962). (Corbis)*

editor-in-chief of Fish and Wildlife Service publications. Her first book, a description of marine life entitled *Under the Sea Wind*, appeared just before the attack on Pearl Harbor in 1941, and therefore went virtually unnoticed. She published a second book on the sea in 1951, *The Sea Around Us*, and it became a best-seller. In 1952, she left her government position and became a full-time writer. Her third book on the sea, *The Edge of the Sea*, appeared in 1955 and was also a great success. Given her deep familiarity with a wide range of wildlife topics from her years with the Fish and Wildlife Service and her reputation as a nature writer, Carson remained watchful for opportunities to explore topics that would enlighten the reading public. She considered writing a book on radioactive fallout from nuclear tests in the Pacific, an issue she believed few people understood well enough to recognize the risks involved. She also signed a contract to write a book about evolution for a science series. By 1957, she had not begun either book and her attention turned back to the issue of DDT.

As an independent scientist and writer, Carson believed that she ought to

Rachel Carson (1907–1964)

Widely celebrated as one of the most significant women in the history of science, Rachel Carson became an icon for environmentalism after her 1962 book, *Silent Spring*, captured the attention of scientists, industry, and the federal government. She intended the book as a call to action, an alert against the dangers of indiscriminate pesticide use, and she based her argument on carefully conducted research of the existing literature on the chemical and biological effects of pesticides. Carson synthesized in *Silent Spring* an argument against DDT that proved as effective as the pesticide was against insects.

Carson grew up in Pennsylvania and from an early age demonstrated a talent for telling stories and writing that foretold a career as an author. She developed an equally avid interest in the natural world. As a student at what is now Chatham College, Carson pursued both her love of writing and her love of nature, but believed she had to choose between them. She chose zoology in her junior year, but she never abandoned

Rachel Carson, shown here giving testimony before Congress in 1963, was a noted biologist and ecology writer whose books played a major role in launching the modern environmental movement. Carson's book Silent Spring, *published in 1962, became a best-seller and touched off a controversy that led to a fundamental shift in the public's attitudes toward the use of pesticides. (Library of Congress)*

devote herself to a project that most closely reflected her devotion to nature. Her focus on pesticides was fixed when three unrelated events came to her attention in a short span of time. The Department of Agriculture used massive applications of dieldrin and heptachlor to kill fire ants, a species despised by farmers and the public alike. According to local reports, the pesticides harmed wildlife and conservation groups criticized the government's use of it. At about the same time, a friend contacted Carson for help in trying to stop DDT spraying for mosquito control in Massachusetts. The friend lived near a bird sanctuary and had found dead birds following previous applications of the pesticide. DDT was mixed with fuel oil to make it adhere to leaves and sprayed from airplanes over broad areas. As Carson began to look into these episodes, she learned of a court case on Long Island, New York, where Director of the American Museum of Natural History Robert Cushman Murphy had pursued legal action against the federal government to stop spraying. From this court case, she obtained a wealth of information implicating pesticide use in the destruc-

writing and, in a sense, integrated her talents in a way that made them inseparable. After earning a masters degree in zoology at Johns Hopkins University, she took a position as a junior aquatic biologist for the U.S. government. Working in the Fish and Wildlife Service, she embarked on a career as a government biologist, but her responsibilities included a great deal of writing and editing. As such, her academic publications as a scientist were limited, but the work enabled her to become a more prolific writer than most of her academic counterparts. Carson wrote three books about the ocean, which she studied professionally as a biologist and passionately as a resident of the Atlantic Coast in Maryland, Massachusetts, and Maine. *Under the Sea Wind* (1941), *The Sea Around Us* (1951), and *The Edge of the Sea* (1955) brought her international acclaim as a writer. More significantly, these books earned the respect of scientific colleagues for the clear and accurate descriptions of marine life.

Carson's writing career arose partly out of the personal necessities of her life. Her father died not long after she completed her courses at Johns Hopkins, leaving her to support her mother in the midst of the Great Depression. Less than a year later, her sister died, leaving two daughters, whom Carson and her mother raised. Her salary as a government biologist barely allowed her to feed this family, so she regularly sent poetry and short articles to various magazines in hope of earning some additional publication income. Persistence paid off. Although her poetry was never published, a nature article would occasionally bring in ten dollars. After the publication of her second book, she quit her job with the Fish and Wildlife Service in 1952 and devoted herself to writing. Although she first considered writing an article on DDT in 1945, she began the project in earnest in 1957. She completed *Silent Spring* in 1962, just eighteen months before she died of cancer.

tion of wildlife. DDT in particular was proving far from harmless to species other than insects.

Throughout the later half of the 1950s, Carson gathered the latest research from colleagues at the Fish and Wildlife Service as well as from researchers elsewhere who had gathered data on changes in wildlife populations. She worked to understand the mechanisms by which DDT moved through the food web, recognizing the similarities between the spread of toxic chemicals and radioactive fallout. These substances were accumulating in the tissues of animals at higher trophic levels as they continued to consume plants and animals that contained even minute amounts of DDT. Her initial plan to write a magazine article or a chapter for a multiauthored book on pesticides gradually evolved into a full-length book of her own. Houghton Mifflin offered her a contract in 1958, and her editor waited patiently for her to complete the research.

Carson's years as a government biologist and writer served her well. She was skilled in gathering material that had been written for highly technical audiences and translating that information into prose that would be readily comprehensible to the general public. Her research was meticulous; she verified every claim she encountered with reputable government or academic scientists. Those who assisted her recognized that Carson needed allies. The chemical companies and government agencies with a stake in widespread pesticide use would not be likely to accept her findings, nor would they be able to profit from the public response that would almost certainly demand changes in current pesticide application practices. And Carson's explicit goal became the public response.

At one point in the project, Houghton Mifflin considered "The Control of Nature" as a working title for the book, but Carson resisted. She acknowledged that pesticide use represented an attempt by humans to control nature, but her focus would be only one of dozens of ways humans attempted to assert control. For a time, the book was to be named "Man Against the Earth," recalling the warlike determination of those who sprayed chemicals on insects and wildlife alike. In the course of correspondence with scientists, bird watchers, and her editors, a number of ideas for book and chapter titles emerged. She received a letter from a woman in Milwaukee who feared that pesticides were responsible for declining bird numbers. From one spring to the next she heard fewer of their songs. Carson's editor imagined a "silent spring" and proposed that phrase as a chapter title. Later, that chapter was retitled "And No Birds Sing," a line from a poem by Keats, and *Silent Spring* became the book's title.

In addition to maintaining a focus on pesticides, Carson also chose to present pesticides as unequivocally harmful to human health and to all living things. As she proceeded with her research, she realized that an argument could be made for restrained use of pesticides in particular circumstances, but she believed that others would make this argument. Her sole purpose was to expose the dangers that currently went unacknowledged by chemical companies and agriculture researchers. Where medical researchers had already uncovered evidence of human health threats, Carson would bring those results to the public.

As Carson prepared her manuscript condemning chemical sprays, research by insect physiologists suggested alternative means of controlling pest populations. Some of this research expanded upon existing methods for killing insects, such as by introducing parasites and predators that would destroy target species. As an alternative to chemical pesticides, Carson hoped this research would complement her argument. Bacteria and viruses that could cause deadly diseases in insects would potentially prove more effective. Most importantly, they would leave no chemical residues that could harm other species. Carson saw even greater promise in other methods of insect population control that had recently been proposed. Hormone disruption of insects might prevent them from maturing and reproducing. Alternatively, male insects might be treated with radioactive cobalt, rendering them sterile. Releasing them in the wild to mate would supplant viable males and lead to a population collapse. The potential success of these methods awaited testing, but these alternatives to DDT advanced her argument in the face of critics who insisted that pesticides were necessary.

Before tackling that argument, Carson chose to open the book with a story that would engage her primary audience, the average American reader. "A Fable for Tomorrow" told the tale of a small town that had once lived in harmony with its natural surroundings. In time, "a strange blight crept over the land and everything began to change" (*Silent Spring*, 2). Farm animals died, songbirds deserted the backyard feeders, and fish disappeared from the streams. The suggestion of an evil spell cast on the town made the story mysterious and its reality doubtful, but Carson closed the chapter with the suggestion that this fable already contained elements of a reality coming into existence in towns across the country. Adding to the tragic consequences, no evil spell had been cast on this town. Carson concluded that the people had done it to themselves, and her book would try to explain what was happening.

In the chapters that followed, Carson explained the means by which DDT moves throughout the environment, killing insects as intended and ending up in water supplies, soil, and animal feed. All of these pathways led back to humans. Condemning the use of DDT on the one hand, she did not suppose that the Department of Agriculture had gone completely mad in supporting programs to spray it across the country. Instead, she sought a more thoughtful approach to insect control, one that looked beyond the imagined devastation of pests that could increase unchecked. She wrote, "All this is not to say there is no insect problem and no need of control. I am saying, rather, that control must be geared to realities, not to mythical situations, and that the methods employed must be such that they do not destroy us along with the insects" (*Silent Spring*, 9). Carson also admitted that chemical pesticides had their place in modern agriculture, even if their recent use had too often been indiscriminate. Her critics recognized no such consideration in her argument. Indeed, few of them read beyond the first chapter, dismissing the entire book as a well-crafted fable.

Carson backed up her argument with figures describing the scope of the problem. For example, in 1947 the government and a limited number of private users had applied nearly 125 million pounds of pesticides, but by 1960, the figure had jumped to over 630 million pounds. She explained how a minute exposure could multiply in the fatty tissues of humans and other animals. As little as one-tenth of a part per million could accumulate and be stored in concentrations as high as ten or fifteen parts per million. An increase of a hundredfold or more could soon lead to illnesses. In animal experiments, concentrations of three parts per million affected the functioning of the heart muscle by inhibiting an essential enzyme and five parts per million initiated liver disintegration. Food and Drug Administration scientists had suggested that DDT posed a hazard to human health that had almost surely been underestimated. Those reports went back to 1950, but the public only first heard of them in *Silent Spring*.

In addition to DDT, Carson described the dangers of a range of insecticides. For each, she indicated where risks were even greater than for DDT and where they were less. She discussed in detail how *herbicides*, chemicals used to control weeds, also posed broader threats to other plants as well as animals and humans. Carson challenged the "legend" that herbicides only affected plants, arguing: "Some are general poisons, some are powerful stimulants of metabolism, causing a fatal rise in body temperature, some induce malignant tumors either alone or in partnership with other chemicals, some strike at the genetic material of the race by causing gene mutations" (*Silent Spring*, 35). She did not, in such cases, distinguish the threat to humans from the threat to living tissues in general. The reader could draw various conclusions from this ambiguousness, but Carson felt no reason for restraint in suggesting the hazards that scientists had already identified or strongly suspected. The time had come for warnings in no uncertain terms, since the public would otherwise remain oblivious to the danger.

One of the most effective means of capturing the public's attention and alerting readers to the risks of pesticides involved comparisons to radiation. Although the spread and persistence of radioactive fallout in the environment was a relatively new concept, the public had become familiar with the risks of exposure. Before mentioning a single pesticide, Carson noted that nuclear tests had spread strontium–90 around the globe, a radioisotope with unknown long-term effects. She went on to insist that chemicals designed to kill insects spread in a similar way. Even herbicides, Carson noted, could serve as mutagens, causing genetic mutations in much the same way as radioactive fallout. She could draw on the research of Nobel Prize–winning geneticist H. J. Muller to strengthen the analogy. Chromosomal changes caused by radiation were similar to changes caused by a growing list of chemicals. That list began with mustard gas and by the mid-twentieth century, included molecules similar to those used in most pesticides. She described the cellular processes that enabled chromosomes to accurately duplicate as part of cell division and quoted the eminent evolutionary biologist George Gaylord Simpson's account of those processes that had served living cells for over five hundred million years. The twentieth-century introduction of chemical and radioactive products in order to disrupt those systems represented a frightening interference with life at its most fundamental level.

Many other details from *Silent Spring* illustrated its significance from the early 1960s onward. Carson also created a lasting change in the relationship between scientists and concern for the environment. By relying heavily on the research of ecologists, wildlife biologists, chemists, physiologists, entomologists, and ornithologists, she established a case against pesticides that proved unassailable. At the same time, pesticide manufacturers and agricultural scientists criticized her for attacking a class of chemicals upon which a large portion of the agricultural economy depended. The conflict over what the scientific evidence suggested depended on two fundamentally different value systems. Valuing the lives of songbirds and humans meant reducing or even eliminating pesticides at the expense of crop productivity, since insects would then regain their position as consumers of products. Since Carson's argument persuaded such a large cross-section of the American public that the risks of pesticides were not worth the benefits to agriculture, the U.S. government was compelled to respond. The Kennedy White House, Congress, and the courts each took unprecedented steps in examining how policies could protect their constituents from environmental harm. While not all of these steps can be linked directly to Carson, over the course of the next decade her work was cited regularly as having set the tone for scientific and public debate about the environment.

Silent Spring made it possible for other scientists to write for a popular audience with greater authority and increased potential for influencing public opinion and the political process. While that book set high standards for writing about the natural world with clarity and grace, others chose to make the attempt on behalf of environmental causes on scales far broader than Carson would have imagined.

Beyond Ecosystems: Systems Ecology

Not all developments in ecological science exposed threats to human health. Academic ecologists pursued a wide range of questions that would have seemed irrelevant to readers of Carson's poetic and provocative work. At Oak Ridge National Laboratory, for example, a group of scientists worked toward a theoretical understanding of ecosystems, including Jerry Olson, George Van Dyne, and Bernard Patten. These three scientists studied ecosystems and succession, as had previous generations of ecologists, but they focused intently on quantitative aspects of the systems. During the 1960s, Olson, Van Dyne, and Patten published several papers that described the transfer of nutrients in ecosystems using differential equations, tracing the nutrients as radioisotopes. They began to call this approach *systems ecology* because their work had grown so distant from the qualitative and descriptive zoological and botanical work of Clements, Shelford, and even Lindeman.

Systems ecology required extensive knowledge of ecosystems as well as a willingness to convert that knowledge to numbers. By manipulating those numbers as mathematical equations, systems ecologists believed they could explore the natural world in a way that would not be possible experimentaliy. Their equations allowed them to make predictions about changes in an ecosystem that might result from alterations in its size or species composition. Without actually destroying habitat, adding species, or shifting nutrient levels, models containing those predictions offered insights into how nature would respond to human activities. Most systems ecologists were not particularly interested in the direct application of their models to management of resources or planning land use development. George Van Dyne, for example, received almost a half million dollars in 1969 to study grasslands, but he did not conduct his study to improve agriculture or even to protect grasslands from human use. Instead, he and others recognized that, in a limited way, their equations represented a means of understanding the workings of the natural world. They also acknowledged that although for the present time they were limited by their ability to solve only certain kinds of equations, computers might eventually revolutionize their ability to model ever more complex systems in nature.

Bernard Patten studied zoology and botany before entering the Army in the 1950s. He had taken almost no mathematics when he started doing research on pesticides as part of his military service. When he realized that his deficiencies in mathematics made it nearly impossible for him to understand new discussions of systems theory as it applied to biology, Patten became intrigued by this quantitative approach. He read widely in other fields and completed his Ph.D. with a dissertation that applied what he found in the broader study of systems theory to a phytoplankton community. When he joined Van Dyne and Jerry Olson at Oak Ridge, their hopes of developing an innovative approach in ecology grew.

Olson arrived at Oak Ridge in 1960 with a Ph.D. in botany from the University of Chicago. Like many of his predecessors at Chicago, he conducted research on dune succession, but in his new post at the national laboratory, his interests turned to the use of radioactive tracers in the study of soil conditions. He began to explore how the analog computers at the laboratory might serve to simulate an ecosystem. The combined interests of Olson, Van Dyne, and Patten launched the most enthusiastic effort in systems ecology with funding from the Ford Foundation and the National Science Foundation. In addition to their research, they promoted systems ecology by teaching a course through the University of Tennessee. Graduate students, postdoctoral researchers, and visiting scientists participated in the course and recognized Oak Ridge as the center of systems ecology. Even after Patten and Van Dyne left the laboratory in 1968, its reputation was established and Olson continued to promote interest in the field there.

Systems ecology attracted scientists in the 1960s and beyond with a promise to explain complex, large-scale biological communities. While ecosystem ecology itself had advanced scientists' methods of recognizing and even quantifying relationships among species, the potential predictive power of models developed in systems ecology proved even more enticing. Previous generations of ecologists often complained that the mathematics only gave the "illusion of accuracy" to claims about how a complex community functioned (McIntosh, 212). Systems ecologists, however, hoped that their approach might yield more robust descriptions of those communities, especially through the use of analog and digital computers. Computer systems themselves became provocative models for how factors within a community might interact.

Tragedy of the Commons

In 1968, ecologist Garrett Hardin articulated the root cause of a particular class of environmental problems. This class of problems, which Hardin called the *tragedy of the commons*, in an article by the same title (1968), included those cases where resource users shared access to the benefits of the resource. All users could take as much of the resource as they liked. Without any incentive to leave a little of the resource for others, it would eventually be consumed and all users would be deprived of future access. The only incentive in such cases was to take as much as possible before the resource disappeared. Many examples illustrate the tragedy of the commons, but Hardin chose the classic example of an unfenced pasture.

Hardin explained that several farmers might own cattle near a common grazing space. Each farmer might bring his cows to the commons, and doing so would not cost the farmer anything. If only a few farmers brought a small number of cattle to the commons, the natural growth of the grass in the pasture could keep pace with their grazing. Since there would be no cost, however, more than one farmer might decide to use the commons as a pasture for cattle. Each farmer would view his actions as economically sound and communally responsible; none would intend to deprive any other of the benefits of the commons. In time, depending on the number of cattle brought by the farmers, the pasture would be unable to provide adequate food for all the cattle. As grass supplies ran short but still provided food for a cow or two, no farmer would have any incentive to start paying for hay to feed his cattle. In fact, each farmer might consider it his right to use the last of the grass. Very soon, the grass would be gone and all the farmers would be forced to buy food for their herds of cattle.

In the example of the cow pasture as a commons, one might argue that the

Boats in the placid waters of the Wahweap Marina at Lake Powell, Arizona. Lake Powell was created by the Glen Canyon dam, which filled the Glen Canyon with water. (Corel)

cattle had used the resource more or less equitably, and that once the grass was gone it could hardly be argued that the resource had been wasted. Hardin, in using such an example, argued that the loss of the commons might be greater than the loss of the value of a particular resource. By the late 1960s, he did not need to imagine the loss of canyons behind a hydroelectric dam and the loss of aquatic life in Lake Erie. Users had destroyed these commons: in the former example the dam supplied future users with electricity and in the latter example the lake served as a waste disposal system. Pollution created a tragedy of the commons, according to Hardin, not by "taking something out of the commons, but [by] putting something in—sewage, or chemical, radioactive, and heat wastes into water . . ." ("Tragedy of the Commons," 1245). Forests cleared would not return to their former state. Wilderness areas despoiled by mining lost the essence of what had made them unique. In such cases, it had proven more economically feasible and communally responsible to change the commons in the short term. Hardin intended to show that the short-term use of a commons generally led to a long-term tragedy.

The future that Hardin imagined included commons destroyed one after another by short-sighted economic gains. He believed that the alternative to these tragedies could be achieved by shifting the incentives. He was also aware, however, that greater incentives would have to be offered in place of the economic gains from the exploitation of the commons. Hardin could not imagine how even a wealthy government could offer adequate incentives. Private enterprise since the time of economist Adam Smith assumed that individual gains would promote the public interest. Hardin questioned the basis of that assumption in examining environmental degradation. In a sense, all of the land, water, and air resources that became polluted in the process of short-term profit making were extracting wealth from the commons and contributing to the tragedy.

The second half of Hardin's article focused on the force that drove the tragedy of the commons. He unequivocally stated that human population created the dilemma. Growing demand for resources, waste disposal, and even for access to national parks arose out of an increasing population. The United Nations had recently declared family size a decision best left to the family itself, and therefore a universal human right. Such a declaration, according to Hardin, went too far by separating human rights from the biological facts. He believed that the tragedy of the commons could not be solved by science or technology. As with the arms race, he suggested that the only solution would involve international cooperation directed toward limiting certain freedoms. Placing limits on reproduction, while generally considered morally reprehensible, would offer relief from future environmental disasters. Just as society imprisoned a bank robber for treating the contents of a bank as a commons, so would society have to treat those who exceeded their fair contribution to future population size. Hardin recognized that many critics would oppose such a limit to reproductive freedom. They might call it coercion. He preferred coercion, however, to the tragedy of the commons.

Hardin's relatively simple model became exceedingly influential, first presented to a division of the American Association for the Advancement of Science, then published in *Science* and reprinted and referred to widely in the popular press. His argument suggested to many would-be environmentalists that science could offer no real hope in preventing or solving environmental problems associated with resource extraction. In an important sense, the futility of trying to protect the commons turned environmentalists from scientific to political solutions. If protection could come only from a major reorientation of economic incentives, then, for the present, searching for additional scientific solutions seemed pointless.

Population Is the Problem

Another ecological voice in the late 1960s pointed more extensively to population as the root of environmental problems. Paul Ehrlich wrote *The Population*

Paul Ehrlich (b. 1932)

While studying insects in the Arctic, Paul Ehrlich began establishing a scientific reputation that has since withstood decades of criticism. As an entomologist and ecologist, Ehrlich, along with his wife Anne Ehrlich, published groundbreaking research that connected the worlds of plant and animal ecology. Before the 1960s, ecologists tended to focus on plants or animals, even when conducting community and ecosystem studies. The Ehrlichs shifted the focus to the relationships between plants and animals in meaningful new ways. Together, they also emphasized the sharing of ecological and scientific knowledge with the general public.

Stanford University Professor of Biology Paul R. Ehrlich helped launch overpopulation as a major concern of the environmental movement with a book he coauthored with his wife, Anne Ehrlich, The Population Bomb *(1968). (Corbis)*

With his reputation firmly established, Paul Ehrlich also became a leading voice in describing the potential dangers of human overpopulation. After a trip to India in 1966, when the reality of hunger and desperation struck him deeply in that overpopulated nation, he began writing about the possibility of worldwide famine. He wrote *The Population Bomb* (1968) to describe the need for limits on human population growth. He advocated zero population growth, fearing that starvation would grip the entire planet within a decade. Despite his warnings, human population continues to grow, and the famine he predicted has not come. Recognizing human ability to defer the consequences of overpopulation, Ehrlich has maintained his argument that the consumption of resources will eventually leave humanity in a state of famine at some point in the future, but he no longer attempts to pinpoint that date.

In the meantime, Ehrlich remains a highly respected scientist at Stanford University. He was awarded a MacArthur Fellowship for his population ecology and conservation work. Along with E.O. Wilson, he is an important advocate for biological diversity and regularly points out the consequences of diminishing that diversity worldwide. He shared the top prize in ecology, the Crafoord Prize, with Wilson in 1990.

Bomb as an explicit call to action, describing the sources of human population growth and the likely outcome of continued increases. With recent data and disasters to support his pessimism, he challenged readers to consider the likelihood of improvements in the human condition. Unless people and their governments implemented dramatic changes in attitudes toward reproduction worldwide, Ehrlich predicted massive famines, disease, war, and suffering.

As an ecologist, Ehrlich described the self-inflicted devastation of humanity with authority. Even more than Rachel Carson, he had established his scientific credentials as an academic scientist and advisor. When Ehrlich published his single-minded criticism of attitudes toward population, he meant to include a wide range of environmental problems that might be solved by a revolutionary shift in those attitudes. Significantly, the direct and indirect effects of DDT were on that list. Like Carson, he pointed to the many ways that pesticides disrupted ecosystems and ultimately damaged human health. In addition, Ehrlich noted that the destruction of mosquito populations by DDT had nearly wiped out malaria in certain tropical areas of the world. That intended and much-praised outcome had the additional effect of reducing the death rate in areas where people already suffered from undernourishment and malnutrition caused by resource shortages in overpopulated regions. Reducing death rates, a central goal of human progress, figured into a population equation that would light the fuse of Ehrlich's population bomb. The other part of that equation, increasing birth rates, provided "the gunpowder for the population explosion" (*Population Bomb*, 28).

Put simply, Ehrlich believed that the impending crisis in numbers resulted from basic features of biology. Humans faced a time when progress created an inescapable bind. The approaching disaster resulted from the culmination of a three-part story that involved "the process of natural selection, the development of culture, and man's swollen head" (*Population Bomb*, 28). Natural selection favored individuals in the human ancestry who reproduced successfully. Those who died without mating or giving birth did not pass their genes on to the next generation. As a result, reproductive rates exceeding the replacement level were both possible and real in the human species. Women could give birth to more then two children in a lifetime, more than replacing themselves and their mates.

The second part of Ehrlich's history of humanity, the development of culture, provided a steady food supply through agriculture over 8,000 years ago. More recently, human culture had included the refinement of medical care. Medicines in the twentieth century enabled people to live considerably longer than their immediate ancestors. Wiping out malaria in certain areas barely compared to the greater reductions in death rates stimulated by modern antibiotics, vaccines, and surgical treatments.

The third part of human history that contributed to Ehrlich's sense of disaster combined biology and culture. He used the phrase, "man's swollen head," to carry a double meaning. A swollen head might suggest pride and hubris, but in this case Ehrlich also meant the literal increase in brain and skull size that had corresponded with the increased period of dependence of children on their parents. Human babies are born with a large head in proportion to their body, but in developing toward adulthood, children require intensive attention and nurturing of mother and father. Such a long period of dependence is rarely found in other species. Ehrlich noted that in humans, part of the assurance that both parents would remain to support children arose from the unique role of sex in reproduction. He wrote: "How could a mother defend and care for her infant during its unusually long period of helplessness? She couldn't, unless Papa hung around. The girls are still working on that problem, but an essential step was to get rid of the short, well-defined breeding season characteristic of most mammals" (*Population Bomb*, 31). To Ehrlich, the human family system, a product of biology and culture, contributed to the intensifying environmental problem of population.

The global community had responded to the growing population by increasing food production. In the mid-1960s, however, a few years of unprecedented agricultural shortfalls signaled to Ehrlich a dangerous turning point. While agriculture had managed to keep pace with population, despite the warnings of British political economist Thomas Malthus almost two centuries earlier, a limit had finally been reached. By the late 1970s, Ehrlich predicted starvation on a scale that even Malthus could not have imagined: hundreds of millions of deaths. He accompanied speculation about famine with predictions of disease, social uproar, and global war.

As an alternative, Ehrlich argued that immediate action by governments, religious organizations, and individuals could prevent disaster. Worldwide educational programs could demonstrate the need to limit reproduction to fewer than two children per adult couple. Government agencies would provide incentives. The aims of "family planning" would shift from helping affluent couples choose when and whether to have three or seven children to helping every couple prevent the conception of more than two children. Ehrlich criticized current programs that failed to acknowledge the global situation. He rejected the optimism of predictions that food supplies would increase once again and become sufficient to meet expanding needs worldwide. He described the worst conditions of currently overpopulated cities and extrapolated to predict a future where all of humanity lived in hunger and despair.

In the end, Ehrlich acknowledged that he might be wrong about what would happen in the coming decade or two. As a scientist, he noted, he lived with the constant potential that his predictions might be disproven. Despite that possibility, he insisted that the fact that half the world's current population lived in misery should provide ample stimulus to adopt his recommendations to limit further growth. He concluded that the current suffering "should be enough to galvanize us into action, regardless of the exact dimensions of the future disaster now staring *Homo sapiens* in the face" (*Population Bomb*, 198). In subsequent discussions of the extent of that disaster, Ehrlich refused to back down, always pointing to additional dangers that could still tip the balance toward global calamity.

A Shifting Role for Science

Far more optimistic than Hardin or Ehrlich, biologist Barry Commoner began a book in the late 1960s that would reveal the many connections between humans, nature, and technology. In *The Closing Circle*, published in 1971, Commoner laid out case after case of the dangerous changes created by humans that ecologists had only recently begun to recognize. Where Carson identified DDT as a single toxic agent that could affect many species, Commoner identified humans as a single species that was, in essence, toxic to all of nature. Despite the destructive potential of humans, a growing general awareness and willingness to respond to environmental problems provided Commoner with hope. He also envisioned a unification of scientific, political, and economic solutions.

Commoner's account began with Earth Day, a nationwide call to action in April 1970, when college campuses came alive with demonstrations pointing out the dangers of environmental destruction. The excitement, however, was not limited to college students. Younger children participated in community cleanups, gathering garbage around schools and parks. Commoner recalled that "determined citizens recaptured the streets from the automobile, at least for a day" (*Closing Circle*, 5). He quoted politicians and government administrators on the seriousness of the movement and its potential significance for lasting political change. Social commentators and activists looked for the source of environmental degradation in capitalism, Christianity, ignorance, selfishness, and innate aggression. The news media caught the spirit of this outcry, helping to point fingers of blame and calling for cures to environmental problems. Among the voices, Commoner cited several biologists, who weighed in on issues ranging from overpopulation to poverty.

The involvement of scientists might have strengthened the calls for action and reform, yet Commoner offered a different assessment. He believed that such universal concern about environmental problems could only emerge from an incomplete understanding of the causes of those problems. The almost unanimous enthusiasm actually reflected the complexity and ambiguity of these issues, which enabled people to "read into it whatever conclusions their own beliefs—about human nature, economics, and politics—suggested" (*Closing Circle*, 10). Critics of a particular economic or political system could blame environmental decay on the excesses of that system, while their opponents played

Barry Commoner (b. 1917)

Prior to the 1960s, few scientists devoted themselves to social and political concerns. Barry Commoner became one of the first to do so, largely in response to the threat of radioactive fallout, which he recognized before most people were even aware that fallout affected human populations around the globe. His efforts to identify the threat of radiation, organize other scientists, and alert the public helped lead the U.S. Senate to confirm the Limited Nuclear Test Ban Treaty in 1963. The treaty ended above ground nuclear tests, which put radiation directly into the atmosphere and caused it to spread widely. Both the U.S. and the Soviet Union agreed to the treaty.

Biologist and educator Barry Commoner is popularly known as "the Paul Revere of ecology" because of his warnings in the 1950s about the health risks of atomic testing, overpopulation, and industrialization. (UPI-Bettman/Corbis)

Commoner was born and raised in New York. His social and political activism emerged while he was a graduate student at Harvard, after completing his undergraduate work at Columbia University. Earning an M.A. and Ph.D. in biology at Harvard, Commoner entered science as a specialist with academic credentials. He studied plant physiology as well as the effects of radiation and various chemicals on living tissue. He also worked as an editor for a popular scientific journal, *Science Illustrated*, before becoming a professor of plant physiology at Washington University in St. Louis in 1947. In that position, he joined the Greater St. Louis Committee for Nuclear Information, a group that notified the public of fallout from nuclear tests in Nevada. He believed that scientific work should respond to social needs, and that scientists should serve as advisors to the public.

As concerns about the environment grew, Commoner played an expanding role in providing information to the public. He studied the effects of pesticides, determining more details of their persistence in soil and water as well as the long-term danger to public health. He wrote *Science and Survival* in 1966 and *The Closing Circle* in 1971, which helped establish his reputation as an activist-scientist. He increasingly suggested that the environmental crisis resulted primarily from technology being used only to create profit for petrochemical companies. His political views eventually led him to run as a presidential candidate for the Citizen's Party. Commoner argued that government failed to prevent pollution of the environment, instead regulating pollution after it has been created and released into the environment. At that point, he insisted, the damage could not be reversed. Since 1981, he has directed the Center for the Biology of Natural Systems at Queens College in New York.

the same game in reverse. Commoner suspected that personal convictions provided more of an impetus for Earth Week than objective knowledge. In response, he hoped to provide some adequate objective knowledge of a series of environmental concerns to establish a firmer basis on which to continue the debate. To begin with, he wanted to narrow the meaning of the environmental crisis around natural phenomena based in the science of ecology.

Commoner introduced the term *ecosphere* to mean the range of habitats on the Earth's surface where life could survive. The ecosphere brought every living thing into contact, recognizing the movement of nutrients through the air and water. He then referred to a global ecosystem, which was an extension of the concept from local areas to the planetary scale. All of the energy and nutrients of all the Earth's ecosystems would flow and cycle from land to air to water, from ocean to continent and back again. The global ecosystem placed all environmental concerns in a context of greater significance. With this perspective, Commoner suggested four laws of ecology that used simple statements of logic founded in deeper ecological conclusions. He appealed to the widespread enthusiasm for commenting on the environmental crisis by making the statements easily memorized and repeated, but he explained them with the rigor of a scientist who understood ecology.

Everything is connected to everything else, Commoner asserted, and called this statement the first law of ecology. Having established the ecosphere as a global ecosystem, and recalling that an ecosystem consists of interconnected parts, he emphasized the implications of those interconnections. The effect of one part on another might influence yet another. Over the nearly 5-billion-year history of the Earth, those effects had shaped life on the planet. Like a captain steering a ship by correcting its course with minor turns of the rudder, small adjustments in one part of the global ecosystem created significant shifts in the environment. One point of this analogy was to demonstrate that attempts to control a particular area through rigid constraint of conditions might cause other areas to veer in unexpected directions. A more effective means of keeping natural systems within familiar ranges, according to Commoner, would be to maintain a flexible approach that could accommodate change. He used natural oscillations in predator and prey populations as an example: "When there are many rabbits the lynx prosper; the rising population of lynx increasingly ravages the rabbit population, reducing it; as the [rabbits] become scarce, there is insufficient food to support the now numerous lynx; as the lynx begin to die off, the rabbits are less fiercely hunted and increase in numbers. And so on" (*Closing Circle*, 35). This well known natural oscillation demonstrated the principle as populations maintained their average numbers over long periods of time.

Commoner offered additional examples that drew on his broader experience with natural systems where long-term effects led to devastation. These examples expanded his audience's ability to reason ecologically. He described *eutrophication* of lakes, a process in which oxygen levels drop so low that fish and other aquatic organisms die. The process usually begins, somewhat ironically, when nutrient levels rise, enabling algae and aquatic plants to grow rapidly. Their growth eventually blocks sunlight to water below the surface, causing algae to die quickly. With large amounts of dead and decaying organic matter in the water, bacteria deplete the oxygen supply, causing further death of virtually all members of the aquatic community. In the case of the lynx and rabbit, environmental changes continually shift to maintain the general course of the populations. In the case of the eutrophic lake, the changes feed into increasingly destructive pathways for the ecosystem. Commoner acknowledged that eutrophication generally was initiated by human practices that inadvertently added large amounts of nutrients to a body of water. The connection between those human practices and the eventual death of an aquatic ecosystem demonstrated the first law of ecology just as clearly as the ongoing oscillations of lynx and rabbit. He concluded his discussion of the law that everything is connected to everything else by referring to the magnification of DDT in moving from soil to earthworms to birds.

The second law of ecology restated a basic physical law, that matter cannot be destroyed. Commoner phrased this law as "everything must go somewhere." In ecology, he added, "there is no such thing as 'waste'" (*Closing Circle*, 39). The most obvious example involves the cycling of carbon dioxide, which is expelled from both plants and animals during respiration. Plants, however, use carbon dioxide in the process of photosynthesis, keeping its overall level from increasing in the ecosphere. Furthermore, photosynthesis produces oxygen, which gets used again in respiration. The same can be said for cycles of nitrogen and phosphorous, important components of organic nutrients. At the time Commoner was writing, other materials, such as mercury, went into landfills or incinerators as part of dry cell batteries when they went dead in flashlights or radios. Since everything must go somewhere, the mercury did not disappear. Often, Commoner explained, the mercury would convert to a vapor and travel in air or water. The heavy metal could end up in the sediments at the bottom of a lake where a fish may ingest it. A man, woman, or child might then eat the fish, and the mercury would begin to damage this person's organs. Ecologists like Eugene and Howard Odum had traced these pathways with radioactive isotopes, and Commoner knew that all elements would continue to travel throughout the ecosphere.

In introducing the third law of ecology, Commoner offered an explicit reference to the value of understanding nature. He wrote—with undisguised reference to the 1950s television comedy *Father Knows Best*—that "nature knows best." Commoner acknowledged that many people preferred to believe that humans have an unlimited ability to "improve on nature," but he endeavored to convince his audience that in the latter half of the twentieth century, much more would be gained from a departure from the arrogance of controlling or improving nature. By a crude analogy, poking a pencil tip into the gears of a mechanical wristwatch, he insisted that few improvements could be expected from the efforts of humans to tinker with nature. In most cases, as with the watch, destruction would result. As an example, Commoner noted that radiation could induce genetic mutations, and that mutations could occasionally improve the fitness of an organism in its environment. For the most part, he quickly added, mutations caused harmful changes that typically resulted in the death of the organism before it even fully developed.

In nature, evolutionary changes had adjusted relationships among species within the ecosphere over hundreds of millions of years. To expect improvements from artificial adjustments made by humans, even when extensive research suggested beneficial outcomes, failed to acknowledge the much greater "research and development" of nature itself. Commoner pointed out that, in the field of organic chemistry, more individual molecules of different proteins could be created than would fit in the observable universe. Since far fewer different molecules of proteins existed, he concluded that nature had already done the research in determining which types would be effective in maintaining living organisms. He even suspected that at some point in evolutionary history, some organism might have produced a molecule of DDT, and like any cell that produces a toxic molecule, it died before it could pass on the genetic information for creating this molecule to another generation. As a lesson, then, Commoner believed that the vast majority of synthetic molecules and chemicals that humans could produce would prove to be carcinogenic or otherwise toxic. He proposed that all such molecules should be tested as thoroughly as drugs before being put to use in the environment.

The fourth law of ecology was borrowed explicitly by Commoner from the field of economics. It was, he acknowledged, the principle that most economists might agree best reduced all the knowledge of their field to a single statement: "There is no such thing as a free lunch." Commoner suggested that in ecology, too, this law encompassed the previous laws. Humans could expect that every gain being won, every kind of improvement in humanity's quality of life will require payment at some later date. He concluded, "Payment of this price cannot be avoided; it can only be delayed. The present environmental crisis is a warning that we have delayed nearly too long" (*Closing Circle*, 46). The four laws of ecology, and especially this last one, became a regular component of scientific and public discourse about the environment in the early 1970s. This was not, however, the extent of Commoner's contributions in *The Closing Circle*.

By drawing on a diverse set of examples, much broader than Carson's intensive examination of DDT, Commoner awakened the concerns of people who had missed the relevance of a pesticide threat. He also provided hope that action might shift the tide of environmental problems. Whereas Ehrlich demanded that people needed to have fewer children as a solution to the overriding problem of population, Commoner identified a wider range of problems that could be targeted individually. He reinvigorated activism that had floundered on the technical controversies of the spread of a particular toxic chemical in the environment.

On the topic of radioactivity, Commoner described developments in nuclear weapons and power generation as a single environmental threat. With all nuclear activity administered by the Atomic Energy Commission in the United States, public faith in nuclear safety rested in the hands of one agency. Commoner showed how the belief that radiation could be contained in production plants and test areas had evaporated as early as 1953, when radioactive fallout from tests in Nevada was detected in Troy, New York, by physicists unconnected with atomic energy research. The notion that radiation exposure was harmless or that certain levels could be considered "acceptable" faded with each revelation of the effects of atomic testing. As a result of the growing awareness of nuclear risk, more scientists had joined the ranks of activists who opposed nuclear power. Banning atmospheric tests of nuclear weapons in 1963, the United States and the Soviet Union jointly agreed that the exposure that resulted from such tests was not worth the risk to the citizens of either nation or the world, even in the midst of an ideological war. Commoner also recalled that nuclear power experts had promised to overcome technical obstacles. Such promises had failed, however, because the danger of fallout continued to outweigh the potential advantage of using atomic power ever since electricity-generating nuclear plants became a reality in 1957. Despite the failed promises, Commoner reported that, by 1975, eighty-four plants would be in operation. The risk of these contained units had led state governments and citizen groups to reject standards considered safe by the Atomic Energy Commission.

In successive chapters, Commoner described changes in air and water quality that resulted from pollutants. His message demonstrated the destructiveness of human action across a wide range of activities. Industries polluted the air with tall smokestacks that could distribute particles and gases into the atmosphere to be carried aloft to other regions of the globe. At the same time, simply driving a car could be even more hazardous when the pollutants of one exhaust pipe were multiplied by the millions of cars on the expanding highway systems of a growing metropolis like Los Angeles. What made air pollution so vexing to Commoner was the way the problem insinuated itself in complex and variable ways. For those experts who had grown familiar with singular, cause-and-effect processes of environmental damage, the haze of air pollution presented new challenges for science. For example, the smog over Los Angeles developed when sunlight interacted with oxygen, nitrogen, and organic compounds in unburned gasoline as it exited in the exhaust of automobiles. Reducing this smog initially meant reengineering internal combustion engines, with the unanticipated result that the smog-creating nitrogen oxide was reduced by creating nitrogen dioxide, a toxic compound that, among other things, damaged human lungs and also reduced tomato plant growth. With further refinements to convert nitrogen and oxygen products to harmless compounds, the damage done by leaded gasoline, hydrocarbon emissions, and carbon monoxide became evident. Commoner managed to outline this seemingly endless process of refinement and recognition of new problems.

One theme that emerged from Commoner's analysis of the environmental crisis was the inadequacy of science in correcting or even identifying the many components of the crisis. As a scientist, he had surprisingly little faith in science and technology. According to him, solutions to environmental problems would not regularly be found in scientific knowledge or its applications. He stated that "we are confronted with a task for which modern science is poorly prepared: the analysis of an intrinsically complex system" (*Closing Circle*, 113). While he astutely described the history of research in chemistry, for example, that had led to the escalation of techniques for producing both organic and synthetic molecules, he concluded that chemists would be unlikely to determine the ultimate usefulness and potential hazards of these compounds at large in the environment. He offered detailed descriptions of technologies that had improved agricultural yields; chemical processes replaced organic soaps with synthetic detergents. Science and technology revolutionized textile production; engineering produced automobiles with more power; electric power generation increased; packaging of virtually every kind of product made them more attractive and more easily transportable. Each of these changes benefited from the expansion of scientific knowledge, and none provided any relief to the environment. Commoner called them all failures of science and technology.

Describing the failure of technology in more detail, Commoner blamed much of the environmental crisis on the narrow vision of researchers who developed technology. In essence, they focused on solving isolated problems, such as the level of particular substances in the flow of a water treatment plant. This narrow focus often resulted in the increase of other substances or the reappearance of the problematic substance further downstream as the environment responded to the remaining components of the waste. This *reductionist approach* attracted enthusiastic scientists as well as funding, leaving the larger problems of understanding broad environmental responses as dull, distasteful exercises. Since reductionism also meant specialization and isolation of scientific disciplines, Commoner believed the appropriate response to the environmental crisis would involve more collaboration around problems of importance to survival. He wrote, "In a word, we need to understand science and technology that is *relevant* to the human condition" (*Closing Circle*, 192).

Commoner recognized that the environmental crisis arose in large part from an economic rationale. He did not suppose that problems resulted merely from ignorance or from the schemes of evil developers. Unlike many environmentalists in the early 1970s, he did not ride the wave of conspiracies to point out the faults of a particular economic, political, or religious ideology as the source of the growing threats. In the final chapters of his book, Commoner reported simply that economic systems place demands on natural resources. The escalations of production in capitalist and communist countries alike, especially since World War II, had incurred a "debt to nature" that resulted from an incentive to add value to goods. Pollution controls and technology that would protect the environment did not add value and had not yet become cost effective in any economic sense. He believed, however, that the day had dawned when environmental protection would begin to drive economic systems. He expected the values of society to respond to the abundant warnings outlined in his book. The skeptics, like Paul Ehrlich and Garrett Hardin, would be disappointed but relieved to observe this response in the coming years.

More significant than whether Commoner was right or wrong in his optimism was the shift he represented in the role of scientists. With the dangers of radiation and pesticides, scientists had largely led the effort in increasing public awareness. Scientists' warnings had alerted an ignorant citizenry to the looming and present dangers surrounding them. Commoner himself initiated several campaigns to educate and transform public opinion and policy. By the time he wrote *The Closing Circle*, environmentalists without scientific training raised concerns about a variety of issues based on analogies to known danger, intuition, and pure speculation. Thus, Commoner and other scientists had become sources of expertise to whom environmentalists could turn for advice. At the same time, polluters and government officials who were reluctant to promote change could also seek scientific advice, giving rise to conflicts in expertise from issues introduced by an interested public audience. Commoner himself welcomed such a role, as well as the context of open debate it engendered. Many of his colleagues, however, found their authority assailed from all directions by skeptics with little understanding of the scientific or technical issues and even less awareness of the realities of uncertainty in scientific research. The composition of public trust in science, especially with regard to environmental issues, had shifted.

Conclusion

The work of Rachel Carson, Garrett Hardin, Paul Ehrlich, and Barry Commoner stands out in environmental history for the dramatic effect these scientists had on society and the American political system. Carson's book stimulated President John F. Kennedy to convene a committee to look into the use of pesticides. The result, a Presidential Science Advisory Committee, continues to advise presidents on a wide range of scientific issues. Commoner's efforts contributed to the signing of the first treaty to limit atomic tests. He went on to advise presidential candidates, injecting environmental issues into national political campaigns for the first time in the late 1960s. He even went on to run for president himself.

In contrast to those environmental scientists who crossed over into environmental activism, the work of systems ecologists, earth scientists, and meteorologists kept a certain distance from social issues and political battles, benefiting from large funding programs, such as the Atomic Energy Commission and the National Science Foundation. Big Science had come to environmental science. The contributions of these scientists, however, remained focused on fundamental questions about energy flow and nutrient cycling in natural systems and on the basic structure of the earth itself.

Geologists during this period embraced the concept of plate tectonics, the idea that the Earth's crust was made up of large, shifting sections. Volcanic activity and earthquakes marked the boundaries of these plates. Such a dynamic model of the Earth provided new and rich questions about geological changes but offered little toward an understanding of the rising tide of environmental problems. Meteorologists, likewise, began to study in more detail the large-scale weather patterns initiated by oceanic currents. They joined in the speculation of how the atmosphere might respond to increasing levels of pollution. Carbon dioxide might serve as a greenhouse gas, warming the earth over time. Chemical pollutants from industrial processes and commercial products might thin a protective layer of ozone in the upper atmosphere. Such speculation did not occupy a significant place in the professional community of scientists during this period.

In the decades that followed, the interplay between environmentalism and environmental science increased. On behalf of the average citizen, environmental activists learned more of the fundamentals of environmental science and, in many cases, developed specialized knowledge in areas of human health and specific environmental risk factors. At the same time, environmental scientists found increasing opportunities within government, industry, and non-profit organizations to research environmental health concerns. The growing prominence of these studies increasingly blurred the line between activist and scientist. At the same time, final authority on virtually any environmental topic became a contested space of competing interests.

Bibliographic Essay

Linda J. Lear has written a thorough biography of Rachel Carson's life, *Rachel Carson: Witness to Nature* (New York: Henry Holt, 1997); see also Linda J. Lear, "Rachel Carson's Silent Spring," *Environmental History Review* 17 (Summer 1993), 23–48. An interesting collection of articles analyzing the rhetoric of *Silent Spring* was recently published, edited by Craig Waddell, *And No Birds Sing: Rhetorical Analysis of Rachel Carson's Silent Spring* (Carbondale, Illinois: Southern Illinois University, 2000), which includes Ralph H. Lutts, "Chemical Fallout: *Silent Spring*, Radioactive Fallout, and the Environmental Movement," (17–41). More on the history of pesticide policy changes is provided by Thomas R. Dunlap, *DDT: Scientists, Citizens, and Public Policy* (Princeton: Princeton University, 1981). On systems ecology, see Frank Benjamin Golley, *A History of the Ecosystem Concept in Ecology: More than the Sum of Its Parts* (New Haven, Connecticut: Yale University, 1993). Robert P. McIntosh describes systems ecology and many characteristics of environmental science in this period in *The Background of Ecology: Concept and Theory* (New York: Cambridge University, 1985). Joel B. Hagen provides a very readable account of developments in ecosystem science in *An Entangled Bank: The Origins of Ecosystem Ecology* (New Brunswick, New Jersey: Rutgers University, 1992). Garrett Hardin's essay has been republished many times, but it originally appeared as "The Tragedy of the Commons," *Science* 162:3859 (Dec. 1968), 1243–1248. Further discussion of the implications of Garrett Hardin's tragedy of the commons appears in Gerald Gardner and Paul Stern, *Environmental Problems and Human Behavior*, 2nd ed. (Boston: Pearson Custom, 2002). Additional comments on Barry Commoner, *The Closing Circle: Nature, Man, and Technology* (New York: Alfred A. Knopf, 1971), can be found in Donald Worster, *Nature's Economy: A History of Ecological Ideas*, 2nd ed. (New York: Cambridge University, 1995), and Samuel P. Hays, *Beauty, Health, and Permanence: Environmental Politics in the United States, 1955–1985* (New York: Cambridge University, 1987). Hays also provides the basis for the introduction to this chapter (see pages 52–53). Both Hays and Worster focus more on the social significance of and response to environmentalism, with little comment on the role of science or scientists.

7

Environmental Science on a Global Scale

As the scale of environmental degradation became increasingly apparent during the 1960s and 1970s, more specialists in environmental science found themselves embroiled in controversy. Problems of pollution and habitat destruction expanded from local concerns to global catastrophes. The social and political consequences of those problems required scientists to engage in a wider range of questions than previous generations of scientists had ever imagined. Controversy loomed on all sides.

In one view of scientific controversy, science is a process of finding particular kinds of knowledge depending upon where and how carefully one looks. While science has and does sometimes operate in this way, a more sophisticated view reveals the more complex factors that influence the production of knowledge. After all, scientists need more than funding; individuals require years of specialized training. During those years, they may acquire the techniques and conceptual information that allows them to speak as experts. They also develop attitudes toward their work and toward their notion of what can be known in their field. Scientists with a given specialization recognize certain limitations in their particular approach and they become aware of how specialists in different areas overcome the same limitations while facing other obstacles.

Even before entering the field that becomes her or his expert area, a future scientist possesses at least a preliminary set of attitudes, aspirations, and personal beliefs. As training proceeds, prevailing paradigms of science may challenge certain values. In some cases, the course of a scientist's career may shift in response to a recognition about how one's individual beliefs agree or conflict with expectations for a particular field. A familiar example would be Rachel Carson, whose affinity for the ocean, ability to write persuasively, and responsibilities for her family led her into a government science position where her primary duties included editing. Meanwhile, others with similar skills undoubtedly became researchers in universities or in industry. This suggests only that by the time a scientist acquires expertise in a given field, she or he may have developed attitudes that would not easily be changed by funding incentives alone. When scientists disagree, the question of honesty or personal integrity is rarely at issue. Instead, the perspectives shaped by their specialized expertise may yield true differences of opinion that are equally based in fact.

While such complexity in scientific controversy has always contributed to disagreements, the post–Earth Day environmental era provided more abundant cases for controversy among scientists than ever before. Rather than simply increasing what science knew, scientists in this period "expanded the realm of what was unknown" in ways that presented obvious challenges to science, policymaking, and public attitudes toward nature (Hays, 338). Previous generations of scientists had developed fields like ecology out of broader zoology, botany, and natural history. They organized new concepts around the interrelation of living communities. With help from allied scientists in geology, meteorology, and biogeochemistry, ecologists formed the core of environmental science. More synthesis of diverse knowledge became an obvious strength of these affiliations. At the same time, the greater number of different methods among scientists practicing in distinct fields exposed environmental science to more visible public awareness of controversies as they arose. Rather than focus on the countless political skirmishes that developed from these scientific controversies, an examination of a limited number of those disagreements can illustrate the challenges that emerged in the science and the environment of the 1980s. A useful sample from this period includes a new holistic theory of the environment, importance of biodiversity, and the impact of acid rain.

The Earth Is an Ancient Goddess

Plant ecologist Frederic Clements argued that a community was itself an organism, not merely an assemblage of individual organisms. His views, sometimes taken quite literally, earned him the reputation of taking the philosophical position of *organicism*. In most instances, Clements used the term organism only metaphorically in reference to communities. Another case of the organism metaphor appeared in 1969 and was applied to a larger system when James Lovelock introduced the idea that all life on Earth contributed to the viability of life itself, with the entire planet functioning as a single organism.

Lovelock chose to call his Earth-as-organism idea "Gaia," after the Greek goddess of the Earth. William Golding, author of *Lord of the Flies* and Lovelock's neighbor at the time, suggested the name as an explicit acknowledgement of the Earth's personal identity. Lovelock intended Gaia to help explain how the Earth could support life at all. He noted that the unique composition of the atmosphere

Scientist and inventor James Lovelock began studying the planet Mars in order to gain insight about the ability of Earth to sustain life. His research led him to propose the Gaia Hypothesis, the idea that life had made Earth livable. (George W. Wright/Corbis)

represented the abundance of life as much as it supported that life. His experience working for NASA gave him unassailable credentials as an atmospheric scientist. He had worked in the Jet Propulsion Laboratory in Pasadena, California, during the 1960s and began to study the atmosphere of Mars when NASA sought a definitive answer to the question of whether life had or could ever exist on that planet. To address that question, Lovelock first wanted to know what distinguished the Earth's atmosphere from that of Venus or Mars. Between 95 and 98 percent of the atmospheres on those planets consist of carbon dioxide, which composes less than one percent of Earth's atmosphere. In addition, Venus and Mars have only traces of oxygen, while that element makes up nearly 21 percent of the air surrounding Earth. Lovelock developed an instrument for sampling trace elements in the earth's atmosphere, a key to establishing the precise amounts of different gases. Even without a final answer to the question of life on Mars, Lovelock made significant contributions to questions of life on Earth.

The elements that compose the atmosphere, mostly nitrogen and oxygen, provide a suitable mix for life to exist. At the same time, without life, the balance of elements would rapidly shift, with oxygen combining into compounds in the Earth's crust. Trees and algae, along with microscopic organisms like diatoms and bacteria, constantly replenish atmospheric oxygen. While animals and all organisms use oxygen for respiration, their activities generally promote a balance and therefore contribute to the production of more atmospheric oxygen by plants. The system, according to Lovelock, maintains life just as it is maintained by life.

The concept of Gaia appealed to Eugene Odum and others who focused on the cycling of nutrients in ecosystems of varying sizes. In addition to elements and nutrients, the Gaia organism could also regulate physical factors like temperature. The Earth as Gaia could reflect or absorb heat from the sun's radiation, and the amount absorbed would vary with the amount and kind of living material on the Earth's surface. Living beings on Earth collectively maintain the carbon dioxide levels in the atmosphere that contribute to the regulation of heat. The sophisticated balance of Gaia provoked strong responses from people who were drawn to its elegance and even its logical simplicity.

Outside the scientific community, those who saw a connection between nature and the supernatural embraced Gaia. They supposed that Lovelock had suggested a mechanism for knowing the role of the divine in maintaining a livable planet for creation. Lovelock himself tried to keep some distance between Gaia and such speculation. His original publication, *Gaia: A New Look at Life on Earth* (1979), made no suggestions about a supernatural regulator. His language, however, did slip in places into constructions that made it seem that Gaia knew what It was doing. In a second book, *The Ages of Gaia: A Biography of Our Living Earth* (1988), Lovelock attempted to state his intentions more clearly for a more popular audience. If he hoped to make less space for God or gods in Gaia in this popular account, he succeeded only in making his ideas more accessible to the audience that found in Gaia an irresistible supernatural explanation for nature's complexity. If Gaia's supernatural elements helped the idea succeed with this audience, those same components made it unsuccessful as an explanatory theory among mainstream scientists. Among those who looked for explanations in observable natural phenomena, Lovelock's Gaia was considered mystical and unscientific.

Lovelock's scientific credibility was not completely compromised, however, due in large part to the contributions of cellular biologist Lynn Margulis. It was Margulis who provided an explanation for how the atmosphere came to have so much oxygen in the first place. She suggested that bacteria had been the first type of organism to develop the ability to convert the sun's radiation into organic sugars—the process of *photosynthesis*. As those bacteria became more numerous, the level of oxygen in the atmosphere rose. Oxygen increased, and more complex organisms, *eukaryotes*, could use the oxygen for more efficient respiration. At a concentration near its present level, oxygen struck a balance between photosynthesizers, respirers, and chemical reactions with elements in the Earth's crust.

Many scientists remained skeptical of Gaia, recalling the criticism Clements faced for his organismic hypotheses. It did not help that spiritual and New Age thinkers had embraced Gaia, but scientists had their own reasons to challenge Lovelock's ideas. The strong version of Gaia, where the Earth maintained a steady state, could not extend indefinitely to the planet's origin, for example. The current steady state provided evidence of a system unlikely to change dramatically. Critics argued that the increase of oxygen in the early atmosphere itself would be at variance with a persistent steady state. They asked why Gaia did not maintain initial conditions. Claims that the planet allowed changes that would provide conditions conducive to the current state were decidedly unscientific and indefensible. One critic of this strong version of Gaia, James W. Kirchner, a geologist at the University of California at Berkeley, also noted that a weaker version of Gaia amounted to nothing new. If Lovelock merely wanted to claim that certain feedback systems helped to regulate nutrient levels and physical conditions like temperature, it would not have been necessary to invent Gaia to do so. Ecology, geology, and biogeochemistry each had established theories of homeostasis that suggest regulatory mechanisms.

Another critic, ecologist Daniel Botkin, did not object to the potential significance of a global model for environmental science. Instead, Botkin saw Gaia as flawed due to Lovelock's inconsistent application of mechanistic and organic metaphors. Since Lovelock suggested that Gaia kept the planet "healthy," it must be like a living thing. In yet another book, *Healing Gaia: Practical Medicine for the Planet* (1991), Lovelock compared environmental remediation to the work of a medical doctor. At the same time, Botkin pointed out, Lovelock likened temperature regulation on the planet to that of an electric oven. As the metaphors became more mixed, Botkin concluded that adherents to Gaia theory would eventually have to succumb to the notion that a purpose or will is responsible for the functioning of the planet. Although few scientists directly deny the existence of God, most, like Botkin, prefer that scientific theories not rely on God as the basis for their explanatory power.

By the late 1980s, the controversy surrounding Gaia had introduced a broad, popular, environmental audience to the idea of a *holistic* organism. Details of Lovelock's proposals remained unclear to most of his non-scientific admirers, but the potential relevance of the message that the Earth takes care of those who care for it extended into the American conscience. Most people also recognized that scientists remained divided over Gaia's accuracy and meaning for science. This degree of uncertainty about a major scientific theory contradicted most people's experience with science as portrayed in popularizations of Newtonian gravity or atomic theory. More like Darwin's claims about evolution, Gaia seemed open to interpretation, depending on one's personal convictions, even among scientists. Lovelock himself added to the uncertainty by noting in places that human activities were merely part of Gaia. People might view pollution and radioactive fallout as factors that might prove harmful to humankind, but such views would make no difference to Gaia. Staunch environmental advocates could not embrace this extension of Lovelock's theorizing.

Among the degrees of uncertainty and contradiction surrounding Gaia, many people found it easy to dismiss the controversy along with the theory. The public often viewed scientists engaged in controversy at this level in one of two ways. First, scientists might be disagreeing over technical details that the average person did not understand. Second, they might be disagreeing on significant points, but the basis for this argument was outside of science in the realm of personal belief or political commitment. Both views of scientists could persist simultaneously. Whatever the case, the public was learning a new form of skepticism about science. This skepticism of science would not remain confined to discussions of Gaia.

Biodiversity as a Global Issue

The extinction of well known, appealing, and economically valuable species in recent centuries has highlighted the peril of nature in the face of human action. Species such as the dodo, great auk, and passenger pigeon became icons of extinction. The American bison, pushed to the brink in the nineteenth century, demonstrated not only how humans could devastate once-great populations, but also how human efforts could restore the remaining fragments of a species. More recently, the California condor joined the ranks of apparent success stories. In the 1980s, scientists recognized a much broader plight relating to species extinction. Rather than the disappearance of charismatic megafauna, ecologists became concerned with habitat destruction that could wipe out entire classes of species and the larger communities that were connected to them. A new worry about diversity among living things, *biodiversity*, took center stage for many ecologists.

Biodiversity connected local, regional, and global conservation efforts unlike any previous incarnation of concern for the environment. As demonstrated by debates over Gaia, scientists increasingly recognized the global interconnectedness of ecological changes and responses. Alterations in the condi-

By the mid-19th century, approximately 15 million buffalo grazed on the Great Plains of the United States. Then the westward expansion of white civilization (such as with the railroad) began to decimate their numbers. Large-scale hunting, often done without any attempt to recover meat from the animals, made them nearly extinct by 1883. The ruthless slaughter also impacted Native Americans, who were reliant on the buffalo, increasing their hostilities toward white settlers. (Library of Congress)

tions of living communities in one area had more obvious effects on surrounding communities and regions. When ecologists and other scientists investigated those effects further, they found deeper implications elsewhere. Local and regional biodiversity appeared as subsets of the global species abundance and variety. Any loss in local biodiversity created a corresponding loss to the planet's living capital. Environmentalists quickly heightened the potential for drama and widespread fear about the consequences of lost biodiversity. The disappearance of a seemingly insignificant worm in a distant forest soil could make news as a lost source of beauty or even a cure for cancer. The media based these claims on the statements of scientists who frankly admitted that they did not know how many species were being lost or how those species might affect humankind.

Beginning in the 1950s, ecologists began developing estimates of diversity. They calculated the number of species in a community based upon the number of individuals found and estimates of the variety or richness expected in a given area. These calculations were aimed at determining the structure of communities, such as what plants and animals provide food for a particular species of animals. The abundance of each species figured significantly in those calculations.

Edward O. Wilson (b. 1929)

As a boy, Edward O. Wilson became enthralled by nature, exploring the swamps around his home in Alabama. An injury left him blind in one eye, and without the benefit of binocular vision, he realized his limitations as a bird watcher. Instead, he turned

his attention to the close-up world of insects. He learned to observe and identify the variety of bugs that crawled across his path, and he made that path wind endlessly through his neighborhood and beyond. When Wilson discovered the Boy Scouts, he realized a way of legitimizing his fascination with nature, earning badges for studying the world around him. He began to study ants in particular, and at age sixteen he initiated a survey of all ant species in the state of Alabama. The task was ambitious for a teenager with no formal training, but he received encouragement from several experts with whom he had already developed a reputation.

World-renowned Harvard biologist and conservationist Edward O. Wilson has written 20 books, won two Pulitzer prizes, and discovered hundreds of new species. Considered to be one of the world's greatest scientists, E.O. Wilson is often called "the father of biodiversity." (Rick Friedman/ Corbis)

When Wilson got serious about his career as an entomologist, he received advice from Philip Jackson Darlington, who was the curator of entomology at the Museum of Comparative Zoology at Harvard University. Darlington told the budding naturalist to do his fieldwork away from the beaten path. Wilson learned never to take the easy path when searching for new species or pursuing rare creatures. As a conse-

Ecologists like Eugene Odum hoped this information would provide insights that would help compare different kinds of familiar locales and discover the structure of less familiar communities. They assumed that the species with the most individuals represented the most important members of the group. By the late 1960s, Odum and others also recognized that a large number of additional species within a community might be represented by only a few individuals. Ecologists compared conditions in polluted and unpolluted habitats, which led to an awareness that unpolluted areas supported both more species and more individuals of all species. Polluted areas had fewer species, each represented by fewer individuals.

Ecologists also developed estimates for the total number of species on Earth, work that went hand in hand with recognizing the potential significance of biodiversity. In 1982, Daniel Botkin and Edward Keller suggested in a textbook

quence of this advice, he began at a young age to discover and catalog species never before encountered by scientists.

Hoping to attend college as a beneficiary of the G.I. Bill, Wilson planned to join the U.S. Army. However, with his injured eye he could not pass the physical exam. Determined to advance his education in pursuit of a career in entomology, he enrolled at the University of Alabama and kept his costs as low as possible. His parents were able to pay the $2,000 it cost for him to earn his degree in three years; he overloaded on classes to save a year's tuition. Wilson began graduate work at the University of Tennessee, but transferred to Harvard where he worked with some of the world's leading entomologists. In Boston, he met and married Renee Kelley in 1955. After earning his Ph.D., he has remained at Harvard throughout his career, becoming a leading entomologist himself.

Wilson's contributions to the science of animal behavior are detailed in his book *Sociobiology: The New Synthesis* (1975), which many scientists in that field consider to be the most influential book of the twentieth century. He connects behaviors from ants to apes as the products of natural selection. Extension of sociobiology to humans created a major controversy in the late 1970s and into the 1980s, and scientists continue to debate Wilson's ideas about the sources of human behavior in evolutionary theory. Before the dust had settled over that controversy, Wilson stepped into another arena of scientific contention, popularizing the word "biodiversity" and leading the fight for preserving large tracts of relatively undisturbed places. This work grew out of his earlier research on island biogeography and abundance of species begun in the 1960s. In 1990, Wilson and Paul Ehrlich shared the Crafoord Prize for their contributions to ecology, an honor equivalent to the Nobel Prize in other fields. He remains an active scientist and scholar and continues to write popular and scientific books about the state of nature and humanity.

entitled *Environmental Studies: The Earth as a Living Planet*—one of the earliest of its kind—that biologists' estimates ranged from three to ten million species, most of which were unknown to science. They explained that although humans had visited and mapped most of the world's surface, biologists who could discover and name species had explored relatively little, so they could not make a precise estimate. Just five years later, entomologist and evolutionary biologist Edward O. Wilson suggested that the number of species of fish, reptiles, birds, and mammals was about 42,580, but went on to explain that this number was insignificant compared with up to 30 million species of invertebrates that scientists had not yet discovered.

Wilson specialized in studies of ants and became the best known voice trumpeting the importance of biodiversity. His work on the subject began as early as 1963, when he and ecologist Robert MacArthur began exploring the notion of species diversity on islands, which they called *island biogeography*. In *The Theory of Island Biogeography*, a textbook on the subject published in 1967, Wilson and MacArthur standardized their theory both mathematically and conceptually for the entire field of ecology. They discussed the South Pacific island of Krakatau as a case study for how species arrive and disperse on an island. Krakatau was a volcanic island that erupted with absolute destructive force in 1883. The eruption killed every living plant and animal on the small fragments of land that remained above sea level, and most of the island collapsed into a massive underwater crater. For months afterwards, the islands on either side of the crater were completely devoid of life. However, over a period of years insects, plants, birds, and eventually reptiles made their way to the barren volcanic islands. Biologists in the twentieth century became interested in how the vegetation and animal life returned and eventually flourished there.

In 1966, Wilson and Daniel Simberloff undertook an experimental approach toward island biogeography that involved creating miniature Krakataus. They began by carefully surveying all of the ant species on ten tiny islands off the coast of Florida. Each island had an area of approximately 1,000 square feet. For example, a circular island this size would have a diameter of forty feet. They found three to seven species of ants on each island. When their surveys were complete, they succeeded in exterminating all creatures on several of those islands by hanging tents entirely over each island and pumping pesticides into the tents. They called it defaunation and considered it an essential first step in tracking the establishment of new communities on the islands. The next step in this experiment was to measure how long it took for ants and other species to come back to the islands, a process referred to as colonization. Factors like distance from the mainland and distance to other inhabited islands were the most significant in determining how quickly populations would reestablish on the islands.

Wilson, Simberloff, and MacArthur succeeded in establishing mathematical models and experimental evidence to support ideas of biogeography that scientists had debated since the time of Darwin. With these models, ecologists could interpret a wide range of data from forest studies in the 1950s. Fragmented forests on the continents went through processes similar to defaunation and colonization of plants and animals on islands. Data from islands from the Caribbean to the South Pacific demonstrated almost identical correlations between island size and number of species. Counting and discovering species became as important to ecologists in the twentieth century as it had been to botanists and zoologists in the seventeenth and eighteenth centuries.

Wilson became a key participant in the "National Forum on BioDiversity," which, in 1988, published a collection of papers on the subject. The volume included articles by Wilson, Paul Ehrlich, Lester Brown, and James Lovelock. These authors expanded their views on ecology and the environment by emphasizing the significance of species in habitats threatened by human activity. Brown, director of the Worldwatch Institute and a world-renowned authority on global environmental issues, argued that the general public needed a broader education regarding the issue of biodiversity. He noted that for too long only a handful of concerned scientists and environmentalists had been aware of the problems facing species in threatened habitats. Lovelock reiterated components of his Gaia hypothesis, emphasizing the role of humans in maintaining the planet as a livable home. He wrote, "I see the world as a living organism of which we are a part; not the owner, nor the tenant, not even a passenger. To exploit such a world on the scale we do is as foolish as it would be to consider our brains supreme and the cells of other organs expendable" (*BioDiversity*, 488–489). Biodiversity became an important unifying concept for ecologists, earth scientists, and environmentalists.

Ten years after the first report, Wilson edited a second volume on biodiversity. In it, he acknowledged the enormous task faced by scientists and global policymakers in preserving diversity and protecting habitats. He emphasized how much people had accomplished in ten years' time, not just by those groups cited by Lester Brown, but also by a wide range of physical scientists, social scientists, geographers, and artists. This last group represented the deeper cultural awareness of biodiversity issues by the mid-1990s.

Wilson continues to write extensively about the need for habitat protection and increased understanding of biodiversity. In his latest book, *The Future of Life* (2002), he linked biodiversity to the health of the planet and the survival of humanity. He relied on estimates of known species both to illustrate the wealth of scientific knowledge and to emphasize how much more there is to be learned. For example, every year botanists identify 2,000 more species of flowering plants, finding up to 60 of these annually in the United States and Canada alone, which are among the most thoroughly examined nations on the planet. As a leading expert on ants, Wilson noted that he and his colleagues know of 10,000 species of ants, but that many more species may exist as yet undiscovered. The most carefully and popularly studied class of species on Earth, the birds, include 10,000 species. That number also may double as remote areas of the tropics are explored more systematically. The rainforests of the tropics are consistently cited as unique treasures with an unsurpassed abundance of species. One hectare in the forests of Brazil, for example, may contain 425 different species of trees. Wilson also points out that the human body is a rainforest of sorts. He notes that "tiny spiderlike mites build nests at the base of your eyelashes. Fungal spores and hyphae on your toenails await the right conditions to sprout a Lilliputian forest. The vast majority of cells in your body are not your own; they belong to bacterial and other microorganismic species" (*Future of Life*, 20). Because those cells are so tiny, he adds, the human body is still mostly human, but every time a person moves about the Earth, she may stir up on the surface of her body seemingly countless individuals of microscopic individuals belonging to species still unknown to science.

As surveys of species diversity continue, scientists raise more questions than they answer. Estimates of the number of species in particular regions of the planet remain uncertain; even carefully explored areas harbor surprises. The collective ambiguity of biodiversity and the science that attempts to describe it creates a challenge for scientists and the public alike. Protecting areas of potentially unique territory, whether large stretches of uninhabited wilderness or small patches of rare habitat, is often costly and politically complicated. Those who favor protection face off against those who favor development for human and economic purposes. Science does not enter these battles with conclusive evidence in support of one side or the other, so once again, the notions of objectivity and the disinterested facts of nature become exposed to the realities of public debate.

Acid Rain as a Regional and Global Concern

By the late 1980s, scientists and policymakers could claim that a consensus had emerged relating to the causes and proper response of at least one controversial environmental issue. Such optimism applied to a phenomenon generally known as *acid rain*. Almost two decades of controversy, alongside the rapid evolution of an environmental regulatory system in the United States, made this consensus remarkable. Acid rain represented a unique form of scientific controversy at the time, a form that has become more common since the 1970s. Part of the uniqueness of this controversy at this time resulted from a blend of scientific uncertainty with growing public caution and limited governmental responsiveness. When scientists first introduced the principles of acid rain, the general public had only recently come to grips with the widespread dangers of pesticides and a limited number of other toxic pollutants. The government had barely established regulations on such pollutants when acid rain emerged as a more diffuse but nevertheless very real threat to the environment. In this case, progress in scientific understanding, public awareness, and government regulation went hand in hand.

Acid rain is the most familiar form of precipitation with elevated acidity content. Scientists refer to snow, sleet, and the less-common dry particulate forms collectively as acid deposition. Awareness of this phenomenon extended

By buffering an acid lake system with 8,000 tons of agricultural lime, Swedish authorities attempt to counteract the continuing inflow of acidifying materials in their waterways. (Ted Spiegel/Corbis)

back into the nineteenth century, when acid in the atmosphere was blamed for corrosion of metals in and around Manchester, England. A hundred years later, a Canadian ecologist, Eville Gorham, began connecting acidification of lakes in England to industrial pollution. Gorham's results, published in the early 1960s, attracted little attention. Although he linked sulfurous emissions from coal and oil burning to a variety of effects in lakes, soil, and even human health, the impact appeared quite local. In addition, while Gorham managed to publish his findings in a variety of journals, they did not attract the special notice of scientists in a given field. The interdisciplinary work needed to conduct these studies made them appear less relevant to readers of specialized journals.

Studies of acid rain faced a major turning point in the late 1960s when fish in Sweden began to die due to acidified conditions. The pollution that caused these fish deaths came from regions far from the occurrences. Acid deposition no longer involved only local concerns. The field of study expanded to a regional and international level of concern. Moreover, Gorham's pioneering interdisciplinary style of research became state of the art in ecological studies of acid rain and its effects.

Lakes in Canada and in the Adirondack Mountains of New York showed similar symptoms of acidification. The source of this acid, according to Gorham and a growing number of scientists, was precipitation falling from an atmosphere polluted by the burning of fossil fuels, which contained sulfur and nitrous oxides. Fossil fuel combustion led to the emission of sulfur in various oxide forms (SO, SO_2, SO_4; collectively referred to as SO_X). Sulfur dioxide (SO_2) was also particularly common. When further oxidized in the atmosphere by reacting with water, smog, ammonia, and other compounds, sulfuric acid (H_2SO_4) resulted. Among the nitrogen oxides, (NO_X), the most common, a single nitrogen atom and oxygen atom bonded together (NO), could oxidize readily in the atmosphere to form nitric acid (HNO_3). The complexity of atmospheric chemistry proved staggering to scientists at first, but with 50 million metric tons of sulfur and nitrogen oxides released per year, the resulting acidity hardly invoked surprise once the reactions were identified.

The enormous quantity of sulfur and nitrous oxide released into the atmosphere each year created local air quality concerns, but regulations in the early 1970s actually sometimes made matters worse. In order to meet local regulatory standards, industries built taller smokestacks on their coal and oil burning plants. Stacks that carried pollutants above 200 feet high ensured that pollution levels in the immediate vicinity would not increase. Those pollutants, however, had to go somewhere. Between 1970 and 1979, industries built an estimated 429 stacks taller than 200 feet high. The vast majority towered over electric utilities. Of these, 178 were more than 500 feet tall, and 36 stood over 800 feet in the sky (Regens and Rycroft, 47). These tall stacks dispersed pollutants to meet local standards, allowing SO_X, NO_X, as well as other compounds to rise high into the atmosphere and travel great distances. Lakes and forests far removed from industrial centers began to feel the effects of coal and oil burning. Scientists found that these compounds traveled several hundred miles a day in cool weather, while SO_4 could persist in upper currents for nearly three weeks when the weather was warm.

Scientists soon linked the transport and eventual deposition of SO_X and NO_X directly to the acidification of lakes in Scandinavia and parts of New England and Canada. Looking first at the *pH scale*, the concentration of hydrogen ions in water, ecologists following Gorham found lakes with readings as low as 2. The pH scale runs from 0 to 14, with low numbers being acidic and high numbers being alkaline. Pure water, sitting at the middle of the scale, 7, has neutral pH. Battery acid has a pH below 1, while lye has a pH near 13. The scale is *logarithmic*, so a solution with a pH of 4 is ten times more acidic than a solution with a pH of 5. At pH 5, a solution is 100 times more acidic than neutral water.

Atmospheric scientists determined that unpolluted rain water actually may have a pH as low as 5. Due to carbon dioxide in the atmosphere reacting with cloud droplets to form carbonic acid (H_2CO_3), rainwater averages a pH of 5.6.

Since this acid is sometimes buffered in soils immediately upon reaching the ground, lakes and rivers can have more neutral and even alkaline pH. When rain with higher acid contents (lower pH) falls on areas without a buffering capacity, there is a drop in the pH of the surface water. Scientists began to find pH as low as 4 in some Adirondack lakes in New York. By 1979, surveys revealed that at least ninety lakes were without fish as a result of acid rain, and more lakes were threatened if change did not come soon.

During the 1980s, scientific studies of acid deposition intensified. While only nineteen reports appeared on the topic between 1973 and 1978, by 1988, scholarly journals had published over 150 more. Scientists quickly became more aware of the connections between air pollution and acid rain. At the same time, their statements on the topic fueled a complex regulatory debate. In the political atmosphere of the 1980s, scientists who described the dangers of air pollution and acid rain were viewed as biased against industry. In attempting to explain the need for regulations in order to protect air quality they lost their scientific objectivity, according to critics. The paradox of experts being unable to provide expert advice without losing their authoritative position led to situations where public officials made judgments contrary to prevailing scientific opinion. In other cases, scientists ended up making policy decisions, but rarely were there scientists in a position to make scientific judgments that would help public officials make policy decisions.

The National Acid Precipitation Assessment Program (NAPAP) originated in 1980 through special appropriations from Congress. Scientists from universities and government agencies around the country toiled throughout the 1980s to gather useful data in the policymaking arena. Their 1990 report pointed to widespread acidification of U.S. lakes. One primary thrust of the report suggested that decreasing pH as a result of industrial pollution would not serve as the only indicator of acid deposition. Instead, *acid neutralizing capacity* provided a more reliable measure of how surface waters changed in response to acid pollution. Focusing on a measure other than pH gave scientists a stronger position from which to argue for regulatory reforms. While pH in surface water could vary for any number of reasons, thus confounding attempts to point out the destructive effect of industry, acid neutralizing capacity suggested deeper patterns. The concept also restored the authority of scientists by putting distance between industry's automatic reaction against regulation and the destructive effects of acidified surface waters like fish death.

Looking solely at lake pH, researchers found that Wisconsin lakes had become less acidic, while detecting no change in New Hampshire, and establishing that only New York's Adirondack mountain lakes had become more acidic. Such changes did not match the evidence of acid precipitation, which showed greater acidity across the continent as prevailing winds moved from west to east. A significant part of the equation for atmospheric chemistry involved more complex factors such as rock and soil types as well as ground and surface water flow. Scientists could compare these factors by examining acid neutralizing capacity, not by surface water pH.

Acid neutralizing capacity results from the interaction of soil with water as it moves into streams and lakes. High limestone content of rock layers helps to neutralize acid precipitation when organic acids break down, forming buffering compounds in the soil. These buffers are more abundant in places like Wisconsin, where limestone makes up large proportions of surface and groundwater retaining soil, but limestone is rarer in the Northeast. In addition, when water flows through buffering soils before entering a lake, the lake has a much greater acid neutralizing capacity than when acidified water flows directly into a lake across the surface. There is less opportunity for this buffering to take place as water flows down steep mountain terrain. As a result, mountains in New York were more acidic than in the low-lying lake region of New Hampshire.

Identifying specific factors that contributed to the acidification of surface waters served an important role as the science of acid rain advanced. Authors of the Assessment Program report recognized the need for regional sampling of water, as well as more extensive data on pH levels and water chemistry prior to the increase in acid rain. Such historical environmental data proved to be difficult to obtain and, in the absence of such evidence, scientists struggled to validate acidification models. Even where data from before 1950 existed, researchers could not readily compare it to more recent data. In the early 1950s, they began measuring surface water pH levels using an electrical device. Earlier data came from measurements using chemical titration methods, which often produced a wider variation in readings. Since the science of the 1970s and 1980s had become increasingly quantitative, and few numeric data points existed from the earlier era, the problem proved especially difficult to set in a historical context. Scientists had no way of demonstrating whether the acidification resulted from new factors in the environment or from new monitoring techniques and recording measures.

While the report called for additional research over wider geographical areas and investigation into the specific chemicals that altered lake and stream pH, the NAPAP authors realized that their best opportunity to recommend effective change would come from direct correlation between prevailing conditions in the most severely acidified areas. Models that focused on soil type and terrain provided the most reliable correlation. In addition, the report pointed to the significant impact of forest changes due to logging, blowdowns, and fire on soils. Deforestation led to increased runoff from soils, limiting the natural buffering of acid precipitation. Agriculture on the other hand, seemed to have little effect on surface water acid neutralizing capacity. While regulators could only address the source of acid precipitation through changes in air quality standards for industry, this report acknowledged that some areas would be harder hit than others and clearly explained why.

The NAPAP report came at the end of over a decade of dispute. Near the end of his term as president in the 1970s, Jimmy Carter hoped to strengthen environmental laws to protect air quality. At the same time, he faced tremendous pressure to find solutions to the perceived energy crisis, a subject that had plagued his administration. He was unable to reach a compromise solution, and the environment took a lower priority. When Ronald Reagan took office as president in 1980, he began to systematically dismantle the federal regulations that many scientists and environmentalists had seen as only a first step. Reagan and his administration believed that smaller government would more effectively improve the economy, paving the way for cleaner industry in the future. The resulting economic growth of the 1980s did not generally translate into environmental improvements since the open market did not replace earlier government incentives to clean up industry. While some polluters did realize the scope of the problem and their contribution to it, only a few of these began to take steps to reduce emissions.

By the time Reagan left office, the scientific evidence linking coal and oil burning to acid deposition had become familiar to scientists and policymakers alike. The public could identify the cause and effect, and President George Bush wasted no time in supporting legislation that would reform existing air quality rules. The 1990 Clean Air Act amendments actually predated the appearance of a final report from the NAPAP, but the media began widely publicizing its findings in the years that led to its finalization. In March 1991, the United States and Canada signed the Air Quality Accord, an agreement to limit emissions of sulfur and nitrous oxides that contributed to acid rain.

Conclusion

By the late 1970s, Americans had seen the dangers of toxic pollution at Love Canal, a chemical dump site on which developers built a residential community just east of Niagara Falls in New York State. They had imagined the threat of nuclear war for decades, juxtaposed with the hope of cheap, clean nuclear energy until the potential dangers of a reactor meltdown became apparent with an accident at Three Mile Island in Pennsylvania. These concerns alone might have prompted deep reforms in the creation of proper disposal of toxic waste and in the generation and consumption of energy. Just as these direct environmental

Cooling towers at Three Mile Island, a nuclear power generating facility in Pennsylvania. A partial reactor meltdown in 1979 demonstrated the potential for technical and human error to cause the release of radioactivity into the environment, even though no significant release occurred on this occasion. (Greenpeace International)

health threats appeared, however, a host of indirect issues arose that seemed to cloud the cause-effect relationships of all sorts of environmental concerns.

The U.S. government responded to environmental degradation with legislation and regulations in the 1960s. In 1970, Congress and Richard Nixon created the Environmental Protection Agency through the National Environmental Protection Act in order to administer regulation and oversee cleanup. Greater authority arrived with passage of the Comprehensive Environmental Response, Compensation, and Liability Act, or CERCLA, commonly known as the Superfund. Identifying polluters, naming them as responsible parties, and compelling them to pay for remediation of contaminated soils, streams, and groundwater became a top priority. The costs of these projects soared, however, and responsible parties sought every possible legal means of escape. In many cases, this involved suggesting reasonable doubt about a company's role in creating the pollution in the first place. In other cases, uncertainty in describing the severity of the problem became an excuse for ignoring it almost entirely.

Scientists examined records, tested soil and water samples, compared data, and often disagreed over what it all meant. Analytical chemists and hydrogeologists could not always combine their expertise to establish a clear picture of what chemicals were present and where and for how long and from what source in a given groundwater system. Troubling as this uncertainty could be, disagreement over facts was less an issue than inability to reconstruct the system from the available facts. With billions of dollars and the economic livelihood of entire communities at stake, courts were reluctant to compel cleanup on the advice of uncertain science. Out-of-court settlements left insufficient funds to effect cleanup. Remediation proceeded at a snail's pace, often out of the public purse. Meanwhile, the hope of new technology that would convert dangerous pollution to valuable energy kept the alchemists' dream alive.

Bibliographic Essay

A comprehensive overview of environmental policies in the post–World War II era is provided by Samuel P. Hays, *Beauty, Health, and Permanence: Environmental Politics in the United States, 1955–1985* (New York: Cambridge University, 1987). Several readable historical accounts of twentieth-century environmental history have appeared recently containing additional social commentary. See J. R. McNeill, *Something New Under the Sun: An Environmental History of the Twentieth-Century World* (New York: W.W. Norton, 2000), and Hal K. Rothman, *The Greening of a Nation? Environmentalism in the United States Since 1945* (Fort Worth, Texas: Harcourt Brace, 1998).

James Lovelock has written a memoir outlining the history of his development of the Gaia hypothesis: *Homage to Gaia: The Life of an Independent Scientist* (New York: Oxford University, 2000). Gaia research updates appear in the *Gaia Circular*, a newsletter of the Society for Research and Education in the Earth System Science.

Edward O. Wilson and Frances M. Peter edited the first report of the National Forum on BioDiversity, simply titled *BioDiversity* (Washington, D.C.: National Academy, 1988); the second report, *Biodiversity II: Understanding and Protecting Our Biological Resources* (Washington, D.C.: Joseph Henry, 1997), followed Wilson's own exploration of the topic, *The Diversity of Life* (New York: W.W. Norton, 1992). In addition, Wilson's memoir, *Naturalist* (Washington, D.C.: Island, 1994), provides details of his life and the foundations of his many contributions to ecological science.

For a detailed analysis of the political debates surrounding acid rain in the United States and Canada, see Leslie R. Alm, *Crossing Borders, Crossing Boundaries: The Role of Scientists in the U.S. and Canada* (Westport, Connecticut: Praeger, 2000). Discussion of the science and early uncertainty about acid rain is provided by Robert H. Boyle and R. Alexander Boyle, *Acid Rain* (New York: Schocken, 1983), and by James L. Regens and Robert W. Rycroft, *The Acid Rain Controversy* (Pittsburgh: University of Pittsburgh, 1988).

8

Environmental Science at the End of Nature

In 1989, journalist Bill McKibben wrote that human society faced the end of nature. In a book on the causes and consequences of *global warming*, he claimed that human activities had altered the environment, making nature now virtually unrecognizable. The 1990s became the warmest decade on record, with seven of the hottest years in recorded history. Changes in global climate, popularly referred to as global warming, corresponded with evidence that human activity affects the weather in lasting ways. While only a handful of scientists made these claims in the late 1980s, environmentalists and journalists like McKibben began connecting the dots. In the years since, data from virtually every field of natural science pointed to persistent climatic changes with significant causes in human industry, transportation, and agriculture.

As evidence of human-induced changes mounted, more environmental scientists joined ranks with environmentalists. With increasing political and social activism among scientists, the cause for concern about the objectivity of science grew. How could science retain its unbiased perspective if scientists adopted environmental views? Raising this question opened new levels of controversy for a much broader range of environmental issues, and scientists became more deeply embroiled in political disagreements. Complex natural phenomena required detailed study, but the results of any one scientific study opened more questions than it answered. Critics on all sides assailed the resulting uncertainty. With each new report, science seemed to lose ground in convincing the public and political leaders. Partly as a result of increasing controversy, the relationship between science and the environment became a political arena in which scientists publicly disagreed over facts and data as well as the interpretations given to those facts and data. Global warming, more than any other issue, illustrated the challenges facing science.

All claims about climate change are based on some combination of observations, mathematical models, culturally significant anecdotes, and memories of past climate. Although many people, including scientists, continue to expect conclusive scientific data to emerge independently, this has occurred rarely in the history of any science. The global climate seems an unlikely candidate for such a demonstration of the incontrovertible wisdom of science. Scientists cannot simply assemble climate data from enough representative points on the planet to state conclusively that climate has changed. Prediction, like weather forecasting, is fraught with complexity and local contingency. Everything from the way scientists frame their questions about the environment to the way the popular media reports results can affect the process of knowing about the natural world.

The science of *global climate change* does, however, resemble other sciences in important ways. Experienced specialists in a variety of subfields make observations. These observations serve as the raw data for mathematical and *scientific models.* As in other fields of science, models imitate the natural world. The simplifications that keep models from becoming as unwieldy as nature itself enable scientists to make comparisons and draw conclusions. In developing climate models that replicate patterns of the planet as a whole, many simplifications become necessary. The resulting models are useful products of rigorous scientific labor. Using them, scientists can develop meaningful scenarios for future warming, cooling, or other changes in the Earth's atmosphere. No single scenario emerges as the predicted planetary future.

To many critics, a science without the capacity for accurate prediction does not seem like much of a science at all. Because scientists necessarily qualify many of their forecasts for the future, a wide segment of society has concluded that global warming is "just a theory," or that scientists have no better idea of what the future holds than anyone else. Many such statements are simply motivated by politics because it may be unpopular or expensive to propose the steps needed to reverse current trends. More significantly, these statements demonstrate the accessibility of weather phenomena to the general public. People spend time talking about the weather in conversation every day. Familiarity with basic ideas about warming and cooling trends, the seasons, and local averages gives almost everyone a sense of expertise when it comes to the weather. Most people recall a particularly hot summer or an icy winter, but such recollections may be the product of only a few hot days in that entire summer, or one big ice storm. Another person might remember a cool, wet week in summer, or relatively balmy winter days filled with sunshine. The individual memory, merely an anecdote in the life of one person, can be suggestive enough to dismiss an extensive set of scientific evidence that demonstrates an increasing trend in hot summers or more severe storm events. Such dismissals of the mounting evidence that human activity has already and will very likely continue to change the climate fail to appreciate the many ways in which science uncovers knowledge about nature. These methods go beyond the simple prediction of likely outcomes.

If the seriousness of global climate change amounted to nothing more than the need for everyone to decide for himself or herself whether the climate is warmer in recent years than in more distant memories, society would indeed be faced with an irresolvable dispute. Contrary to the picture some critics attempt to paint, science provides a highly consistent theoretical and empirical story of climate change. Moreover, in that story the earth is warming, and that warming is significantly influenced by human sources like industry, transportation, and agriculture. By the same token, environmentalists taking a more extreme view of a planet in dire jeopardy are painting a picture that would require a radical restructuring of all human societies, and especially of the industrial capitalist societies like those in the United States, Europe, and Japan. The political choices faced by this and the next generation depend more upon an understanding of what scientists have learned about climate than upon the rhetoric of opposing extremes. So the questions become: What is known about the world's climate and weather, and what kind of future might one expect to see?

What Is Known

The popular media typically refers to global climate change as the greenhouse effect. This phenomenon is so named because of its similarity to how sunlight heats a transparent enclosure for plants. Even with freezing temperatures outside, a greenhouse can remain warm throughout a long winter if the sun shines on it for at least a few hours each day. The warming is the result of radiation in the form of *visible light* passing through the glass. That light warms the surfaces within the greenhouse, which are typically dark in color and designed to reflect very little of the light. Warming those surfaces transforms visible light radiation to *infrared radiation*, which is invisible to the naked eye. Infrared radiation passes through materials such as glass very inefficiently, and as a result, the radiation inside the greenhouse is trapped. The resulting warmth escapes slowly, so that by the time the sun rises the next morning, temperatures are maintained at a high enough level to support plant life in the greenhouse. An even more vivid example of this effect is familiar to anyone who has ever parked a car in a sunny place with all the windows closed. After a relatively short period of time, the car's interior will become unbearably hot. The heat is infrared radiation trapped by the car's windows, which had allowed visible light to pass through and warm up the dashboard and seats.

The Earth's atmosphere, thanks to gases like carbon dioxide, methane, nitrous oxide, and even water vapor, traps infrared radiation. When visible light from the sun warms oceans, forests, parking lots, and playgrounds, those surfaces warm and emit radiation that cannot pass as easily through the atmosphere to return back into space. Without these gases, this planet would be nearly as cold as the moon or Mars. If concentrations of greenhouse gases increased tremendously, Earth might be more like Venus, which has thick clouds of greenhouse gases that trap the sun's energy, maintaining the planet's surface at temperatures that are too high for most life forms.

Greenhouse gases do not directly influence the weather. Weather, by definition, is a local phenomenon resulting from atmospheric conditions that are typically identifiable in advance. While it is virtually impossible to predict the weather with certainty, meteorologists can provide forecasts that are generally reliable for up to forty-eight hours. They do so by comparing conditions at a given time with periods in the past when conditions were similar. Such comparisons allow them to forecast conditions over the ensuing hours based upon what happened in the past under similar conditions. Visual images generated by satellite photographs and computer enhancements also assist in daily or even weekly weather forecasts. Extending such forecasts beyond a few days involves simplifying or ignoring an ever-increasing number of variables, and such extended forecasts typically require revision on a daily basis. Anyone planning an outdoor event who begins to watch forecasts over a week in advance will experience multiple changes of fortune in expecting fair weather to arrive at the appointed time.

Climate differs from *weather* in several respects. First, scientists can describe regional or even global climate. Unlike weather, climate exhibits greater consistency over larger areas of the planet. A tropical climate might include vast reaches near the equator. Temperate climates exist in many areas around the globe, generally exhibiting more dramatic changes of season than tropical or polar climates. Season is a second feature of climate that is different from weather. While weather during different seasons may vary dramatically in a given location, a particular climate may have relatively consistent seasons, year after year. For example, northern temperate regions with cold winters tend to have cold winters every year. The weather on a given day in such a place may be many degrees warmer or colder than the day before. Variations in weather are less significant than variations in climate. Also, scientists define climates by data that is highly statistical. Meteorologists compare temperature, precipitation, elevation above sea level, proximity of land masses or oceans, and composition of the atmosphere. These comparisons enable them to distinguish one climate from another. Tracking changes in any of these factors can lead to a better understanding of what features of a given climate are significant.

In his 1864 book, *Man and Nature*, George Perkins Marsh pondered the effect that deforestation might have on the global balance of carbon. He acknowledged the complex relationship between atmospheric chemistry, *mete-*

orology, and tree growth, but hesitated to speculate on the degree to which the atmosphere might change due to alterations in forest lands. Given the goal of his book—to alert readers to changes being wrought by human activities—Marsh seems an obvious place to start talking about climate change. However, his writing on the subject did not move beyond mere speculation. Svante August Arrhenius, a Swedish chemist who, in the late nineteenth century, suggested that industrial activities could increase the levels of gases in the atmosphere that trap heat, made a more substantial contribution to the history of atmospheric science. In particular, Arrhenius noted that carbon dioxide, a byproduct of many industrial processes and the burning of fossil fuels, could lead to a steady warming. Without any means to measure and monitor carbon dioxide in the atmosphere, the idea attracted little attention.

As the industrial revolution accelerated into the twentieth century, the amount of carbon dioxide released into the atmosphere occasionally aroused the concern of scientists interested in the chemistry of the earth's atmosphere. Before sounding any alarms, they realized that carbon dioxide dissolves in water, and as concentrations increase in the air, the oceans of the world would continually absorb more carbon dioxide. This reasoning seemed to put the issue to rest. By the 1950s, however, Roger Revelle and Hans Suess, working at the Scripps Institution of Oceanography in California, concluded that oceans could absorb carbon dioxide only within certain limits. Beyond that, they acknowledged, atmospheric concentrations would continue to increase.

Perhaps the best known confirmation of carbon dioxide increase comes from the atmospheric observatory on Mauna Loa, on the island of Hawaii. There, Charles Keeling set up an instrument to detect and measure carbon dioxide. In such a remote part of the world, one would expect to find little evidence of industrial activities. Nevertheless, Keeling's data have shown two significant trends in atmospheric carbon dioxide levels. First, the gas has an annual cycle that corresponds with the changing seasons. Concentrations decrease as plant growth withdraws carbon dioxide from the air, and as plants die and drop their vegetation, more carbon dioxide is released. The worldwide cycle repeats every year. Second, among the annual variation, there is a consistent overall trend. Keeling measured atmospheric carbon dioxide on Mauna Loa below 320 parts per million prior to 1960. A steady increase since that time placed it above 350 parts per million by 1990. This data set provides an unmistakable indication of human activities, although the significance of its connection to global climate change continues to be debated.

The importance of Keeling's data becomes more striking when compared with evidence of carbon dioxide levels dating back approximately 1,000 years. Scientists can collect this long-range data from air bubbles trapped in glacial ice

Curator Geoffrey Hargreaves inspects core samples from the Greenland ice sheet. They are stored in a freezer at -33° F. The cores will be examined for evidence of global warming caused by rising CO_2 levels. (Roger Ressmeyer/Corbis)

over Antarctica. Concentrations of carbon dioxide in air from previous centuries are available in layers of ice frozen at regular intervals and buried by additional layers. They check those measurements against other samples for consistency. Ice cores generally reveal carbon dioxide levels below 300 parts per million prior to the nineteenth century.

Carbon dioxide is an enormously significant component of living systems. On the scale of the planetary atmosphere, however, the gas comprises only a small fraction of one percent of the air. Three hundred *parts per million* is three-hundredths of a percent. Gases that are much more abundant are nitrogen (N_2), which accounts for over 78 percent of the atmosphere, and oxygen (O_2), nearly 21 percent. The remaining one percent is mostly argon, an inert gas that does not react with other atmospheric chemical substances. Along with methane (CH_4) and nitrous oxide (N_2O), carbon dioxide brings the total of greenhouse gases in the atmosphere up to less than one-tenth of one percent. The reason these gases contribute so significantly to the Earth's climate has to do with the way the sun's radiation passes through, is reflected, and is absorbed by the atmosphere.

Collecting the data that helps climate scientists define their subject has been an ongoing project for centuries. While the work of naturalists established certain baselines for climate studies, much of what scientists in the late twentieth century hoped to understand depends upon data that became available only very recently. Data from oceanographers and atmospheric chemists like Roger Revelle, Hans Suess, and Charles Keeling are invaluable. Those data are also sometimes difficult to compare to the information collected even a few decades earlier. Trends that emerge once scientists have measured atmospheric levels must be examined for continuity with data that preceded those measurement techniques. Sometimes, as with the case of air bubbles trapped in ice, scientists can make new measurements of past climates. In other cases, scientists are unable to obtain data from the more distant past.

As new technology continues to emerge, scientists make use of more accurate and more complete readings about climate. Satellites collect data that represents a vast improvement over earthbound readings, yet comparing the new data to the old readings presents similar problems of continuity. Oceanographers began amassing measurements aboard dedicated research ships beginning in the middle of the nineteenth century. Recently, buoys with automated measuring and recording equipment collect data in more places and for longer periods of time than any ship and crew. What to make of this additional data, unprecedented for comparison purposes, presents scientists with what sometimes seems like too much of a good thing.

In the early 1990s, the United Nations assembled the Intergovernmental Panel on Climate Change (IPCC), which worked to produce three extensive reports on what scientists knew about the climate. The last of these appeared in 2001 and began with a technical summary for policymakers. The summary was meant to contain the information about which scientists were most certain and to provide a firm basis for reasonable debate over proposed policies that might be aimed at addressing concerns about climate change. It contained very little of the language of uncertainty, which had served as a hallmark of this science. The summary frankly referred to data collected from around the world, encapsulating the data on a few pages with simple graphs.

The graphs illustrating the change in global temperature used the period from 1961 to 1990 as a baseline. Discussion did not focus on whether scientists agreed on the normal temperature of the Earth's atmosphere. Instead, the graphs starkly showed that in the years prior to 1961, temperatures had tended to be lower. A range of temperatures within a half degree Celsius below the 1961–1990 average had persisted for a thousand years or more. Beginning near the turn of the twentieth century, the range began to shift upward. The trend appears to continue upward entering the twenty-first century. Looking at a graph on the scale of the past 150 years, the trend appears to climb steadily. On a graph of the past thousand years, the increase in temperature appears recent and abrupt. The authors included both graphs in the summary for policymakers.

The temperature graphs precede another set of time scale graphs. Showing atmospheric concentrations of three gases—carbon dioxide, methane, and nitrous oxide—the second set of graphs correlate closely to the temperature graphs. All three gases show stable concentrations for nearly 900 years, climbing steadily after about 1850. Those gases each contribute to global warming. In addition, while all three occur naturally, their increase in the past 150 years results from human activity. The burning of fossil fuels, agricultural activity, and industrial processes account for the increase.

The summary for policymakers thus identifies an unmistakable trend toward increasing temperature. The increase is linked to rising concentrations of greenhouse gases. Finally, those rising concentrations result from human activity. The clear implication is that human activities caused increasing global temperatures. The remainder of the report provides details of how scientists have drawn inferences from the data at hand. Those details, however, do little to clarify how the implied problem might be solved.

The Changing Atmosphere

The mere fact that scientists have observed changes in the atmosphere over the past 150 years and that they can collect meaningful records going back more than 1,000 years only hints at the more extensive changes that have occurred in the atmosphere during the history of the planet. That history, stretching back over 4 billion years, includes periods of warming and cooling that exceed current trends. Records of those periods appear in fossilized vegetation and in geological evidence around the world. Scientists began to look for patterns in the warming and cooling events that might point to a cause for the most recent increase in global temperature. This search led to the correlation of factors in the Earth's orbit around the sun that affect global climate. The annual sequence of seasons is caused by the tilt of the Earth in relation to the sun. In summer, the tilt of the northern hemisphere toward the sun provides more direct rays and warmer temperatures. At the same time in the southern hemisphere, the planet receives less direct rays, causing cooler winter conditions there. The situation is reversed during winter in the northern hemisphere. The tilt of the Earth that is responsible for this phenomenon is not perfectly constant. Over periods of tens of thousands of years, the tilt shifts slightly, making summer a longer and warmer season for many centuries at a time, then making winter longer and colder for long periods. The distribution of land and oceans on the planet also means that long winters in the northern hemisphere may lead to accumulating snows that do not melt during the short, cool summers. Periods of persistent snow and ice have been

Scientists have gathered evidence suggesting that glaciers are melting at an increased rate, a sign that global temperatures are on the rise. (National Oceanic & Atmospheric Administration)

known as ice ages ever since Louis Agassiz first recognized the evidence for these periods in the nineteenth century.

The last ice age ended nearly 11,000 years ago, corresponding to a time when the earth's axis was tilted slightly differently than it is now. That tilt had been enough to lengthen the winters in the northern hemisphere. As snow accumulated from year to year, not completely melting during the brief summers, it compacted into ice and began to spread across vast areas of North America, Europe, and Asia. Observers can easily identify records of this ice age in regions of those continents where the glaciers finally began to melt. At those melting boundaries, enormous amounts of sediment were deposited, and level plains were left behind by the meltwater. Similar events at fairly regular intervals in the past correspond to the regular fluctuations of the Earth's orbit around the sun. Some climatic shifts become larger as multiple fluctuations of the orbit coincide, while other shifts are smaller when fluctuations cancel each other out.

Scientists can forecast these global climatic changes well enough to suggest that another ice age can be expected in the next 10,000 to 15,000 years. While this does not provide a useful prediction of what to expect on the scale of human lifetimes, it does identify the continuing variation of the Earth's climate. In the meantime, Earth is experiencing a relatively warm, interglacial period. The last interglacial period occurred about 120,000 years ago. The coldest of the relatively recent glacial periods took place about 2 million years ago. All of these dates suggest climate change on scales of thousands of years or longer. In addition, in no previous case were human activities a part of the factors that led to a change in climate.

The time scales that people grow accustomed to generally fit within a human lifespan: a year, a decade, a twenty-fifth anniversary. Discussion of changes that take place over several centuries or millennia, to say nothing of millions of years, often leave people with a sense of skepticism or discomfort. Many scientists, including geologists and climatologists, work at understanding natural phenomena that regularly take place over these vast stretches of time. Those phenomena often involve changes that seem slight, even on the scale of a human lifetime. A degree or two in temperature hardly seems worthy of serious concern, and stretched over several decades, most people would consider the change imperceptible, except for the detailed observations and record-keeping of science. Realizing this important contribution of science might make it easier to accept the cautious forecasts of climatologists and earth scientists who now suggest a trend that may affect much of the life on this planet.

Perhaps the largest hurdle in accepting that the earth's climate is getting warmer is the difficulty in directly linking changing temperatures to human activities. If global warming is happening, how do scientists know that humans are to blame? The connection here is complex, especially when one considers the many warnings that scientists themselves offer about confusing *correlation* with *causation*. On a regular basis, the public learns that just because two events occur in sequence, one after the other, the first is not necessarily the cause of the second. It may be the case that a different, invisible, or unnoticed factor has caused the first and second observed events. In such a case, the two observed events are correlated, but the first does not cause the second. In other cases, however, the persistent correlation of two events does suggest that the one causes the other.

As an example of how correlation and causation might be emerging in the case of global warming, consider first the analogous history of cigarette smoking. Rates of lung cancer increased dramatically during the twentieth century. Before 1900, the disease was extremely rare. By the 1960s, lung cancer had become one of the top killers in developed countries. Medical doctors and researchers began to consider the corresponding rise in cigarette smoking in those countries where lung cancer had increased. There was no direct evidence that tobacco caused lung cancer. In fact, people had used and smoked tobacco for thousands of years. Yet, the correlation raised suspicions. Soon, other links between tobacco smoke and lung diseases emerged, largely as a result of closer inspection of the possible correlation. In smoking cigarettes, people were inhaling more smoke containing active chemicals than had commonly been the practice throughout history, and their availability made the percentage of people who smoke rise to unprecedented levels. Now, numerous physiological explanations directly connect tobacco smoking to lung cancer. The correlation, however, offered the first clue to this causation. Similarly, scientists are regularly uncovering additional connections between human activities and global warming but, in the meantime, the correlated trends offer significant evidence for consideration.

One important lesson to take from this cautionary tale about correlation and causation suggests that just as human activities might contribute to global warming, changes in human activities might reverse or at least slow the trend. Another recent example from atmospheric science illustrates this lesson. In the 1930s, a group of synthetic chemicals known as *chlorofluorocarbons* (CFCs) became widely available for a number of commercial applications. Originally introduced as refrigerator liquids, CFCs soon became useful as gases that could be stored under pressure and would expand to propel aerosols and to create styrofoam. Tests showed that these gases were nontoxic and would not react with other compounds. Experts considered CFCs absolutely safe. Part of the utility of CFCs, their nonreactive properties, eventually emerged as their greatest threat.

Since CFCs do not react readily, they do not break down easily. These gases remain in the atmosphere, and eventually some quantities rise high above the Earth's surface. In the outer layers of the atmosphere, particularly the stratosphere, the sun's radiation contains enough energy to break CFCs apart. When they begin to break down, they can react with other compounds. At those high elevations, *ozone* is more abundant. It is in the stratosphere, in fact, that ozone blocks many of the sun's high-energy *ultraviolet radiation*, reflecting it back into space. When CFCs break down and react with ozone, less ozone is available to block ultraviolet radiation. When scientists recognized these reactions, efforts to stop the production and use of CFCs began.

In 1987, government representatives from most of the world's industrialized nations met in Montreal, Canada, and agreed to end production and use of the most harmful CFCs. They also agreed to phase out use of the less harmful versions. In just fifteen years, scientists have recorded a reduction in the rate of ozone depletion. This means that although ozone levels in the stratosphere continue to decline, the decline is less rapid and may eventually level out. Atmospheric chemists identified the problem in time to enact societal changes that slowed ozone decline. Nevertheless, increasing levels of ultraviolet radiation today are linked to a rise in skin cancer, especially in the southern hemisphere where ozone levels had dropped most severely. If agreement on global warming could be reached, as it was with CFCs and the ozone, some slowing of the current warming trend might also be possible.

Science and Models of Climate

Convincing governments, the public, and industries to make changes in regulations, purchasing habits, and manufacturing processes is difficult. Any of those changes requires widespread support and a willingness to take on enormous costs. As with the links between cigarette smoking and lung cancer, as well as that between CFCs and ozone depletion, scientists are called upon to provide evidence that changes are necessary and worth the cost. So far, society has been unwilling to accept the scientific evidence currently available, and scientists continue to question that evidence, searching for a more complete understanding of the global climate.

More data will help in providing convincing evidence of global warming. In addition, scientists continue to model climatic changes. They use the additional data collected, along with more powerful computers to simulate different conditions. Creating forecasts based on these data and models involves taking into account a variety of assumptions. One approach considers the climate in the past as an analogue to the climate today. By collecting historical climate data and creating a model that accurately recreates that past climate, scientists can use that model to determine with fair accuracy what may happen to the climate in the future. Such models can use all of the available data, and scientists can add more data for refinement as it is collected. Of course, the major criticism of these analogue models is that the climate of the future may not resemble the climate of the past, for reasons climatologists have not yet recognized. Among those reasons, of course, is the possibility that human activities have already altered some basic assumptions about how the global climate operates.

Computer models provide greater promise for forecasting the future climate. Modelers enter basic physical principles about the earth's system. They describe for the computer the earth's size, land mass, ocean volume, atmospheric components, and certain details about vegetation. They enter into these models the physical laws that describe the motion of particles, energy levels from varying forms of radiation, properties of gases, and effects of different amounts of moisture. Finally, as accurately as possible, the current conditions of the Earth's climate complete the equation and the computer is set to run. Each new generation of computers, with increasing speed and computation abilities, makes the model more realistic.

Climate modelers began in the 1970s with one of two basic approaches.

First, they attempted to model the circulation of the gases in the atmosphere. By focusing on the atmosphere, they hoped to get the best picture of what the climate was like. From these models, they could get three-dimensional images and readings. The second generation of computer models focused on the circulation of the oceans. Because water covers so much of the earth's surface, and because moisture plays such an important role in climate from place to place, the oceanic circulation approach promised to provide new understanding of the climate. Most recently, modelers have combined these two early approaches, aided by the continued increase in available (and affordable) computer power. The "coupled atmospheric-oceanic general circulation models" represent the state of the art models in forecasting possible future climates.

With such powerful and advanced methods for modeling climate, many in the general public expect scientists to offer conclusive predictions about the future. Policymakers also hope for concrete statements about what will happen and what they can do in the meantime. Scientists, however, have built into these models realistic expectations about the wide variations of outcomes that may result from short-term climate scenarios. Given the range of variation in these cases, long-term forecasts become much less clear cut than the general public and the expectant policymakers might hope. By comparing a set of worst-case scenarios with best-case scenarios, modelers can suggest a range of temperature increases. Since no current scenarios forecast stable or decreasing temperatures, planning for some increase would be in everyone's best interest. Policymakers have often taken the range of possible increases to mean that scientists remain uncertain, and even they suggest that uncertainty makes action imprudent at this juncture. The general public sees this conflict between science and politics as a battle over ideologies. What policymakers consider uncertainty in this context deserves more explanation here.

One of the major sources of variation in climate models, which contributes to the so-called uncertainty in climate models, involves what scientists refer to as *feedback*. Some forms of feedback, known as positive feedback, increase the functioning of the system. In the case of the climate and global warming, increased temperatures create more evaporation, putting more water vapor into the atmosphere. Since water vapor is a greenhouse gas, this creates the potential for further warming and can lead to still more evaporation. Other forms of feedback are known as negative feedback. One example of negative feedback in the climate system is that as the lower atmosphere warms, it releases more infrared radiation to the upper atmosphere. Many scientists expect the release of this radiation, or heat, into space to become more efficient as the lower atmosphere becomes warmer. As a result, global warming may be offset in part by increased heat loss into space. Scientists cannot predict exactly how these and other examples of both positive and negative feedback will operate on a global scale over long periods of time. To even provide a reasonable forecast of such processes would require greater experience with the global climate system, at least over the short term. Since no one has access to such experience yet, in their models scientists include a variety of scenarios that should represent the range of possibilities.

How the general public and policymakers interpret the range of possibilities in climate models may be open to debate. Scientists typically recognize the potential for that debate to escalate, but as more information becomes available, they must continue to explore new avenues. For example, several scientific studies support the idea that higher carbon dioxide concentrations in the atmosphere can actually increase plant growth. If plants, especially agricultural crops, would grow more rapidly with added greenhouse gases, the improved productivity of farmland might offset some of the potential disadvantages of global warming. Many agricultural crops, however, derive no advantage from increased carbon dioxide concentrations. Rising temperatures around the world could pose other potential advantages for agriculture in northern zones. Vast areas of Canada and Russia where growing seasons were previously too short and cold might support crops. Those regions might make up for the loss of productive farmland in *tropical* and more *temperate* zones. Plant scientists and climatologists alike have agreed on this possible outcome. Whether such advantages in crop production justify continued carbon dioxide increases, however, is not a decision that scientists can make. Some scientists have touted these advantages in arguing against regulations on greenhouse gases, but it becomes problematic to see the scientist as an objective source of information and expertise in this context. The Intergovernmental Panel on Climate Change (IPCC) generally has not promoted such conclusions. For each set of potential benefits, most scientists see an even larger set of risks, some of which are as yet unforeseen.

In fact, most of the more optimistic forecasts rely on simpler models of the future climate. By including a few negative feedback mechanisms that limit the processes of global warming, and even fewer positive feedback systems that enhance it, modelers can create scenarios that show a stable global climate. The IPCC, however, has repeatedly stated that such scenarios are unlikely. After considering the many ways that scientists and policymakers can interpret evidence and the different assumptions they might put into modeling future climates, the consensus of scientists view continued global warming as the result of human activities dating back to the Industrial Revolution. Since 1995, the IPCC has maintained that "the balance of evidence suggests that there is a discernible human influence on global climate" (*Climate Change 2001*, 22). Challenges to that consensus view continue to attract attention, and as part of the process of science, these challenges continue to face scrutiny from both the scientific community and the general public.

Environmental Problems and Human Behavior

Many observers of the global warming debate over the past two decades have begun to feel that there is no way out of this controversy. Evidence from so many different sets of data is interpreted from equally many viewpoints. Individual experiences and cultural anecdotes about weather continue to cloud attempts to clarify long-term global trends. The global warming debate can be viewed like so many other controversies in science, where politics and personal beliefs make the resolution of the issue virtually impossible. Fairly recently, however, social scientists have made progress into understanding the ways such controversies emerge and continue in broader social settings. Environmental issues represent an entire area of research for sociologists, social psychologists, anthropologists, ethicists, and others. Their findings so far have not provided a clear road out of the controversies faced by modern societies, but they have clarified the issues considerably. Most significantly, social scientists have managed to distinguish more precisely between the components of controversies. In doing so, they have begun to identify the parts of the controversy where scientific evidence can be helpful in providing information to the general public and to policymakers.

Resolving environmental controversies such as global warming involves more than recognizing warning signs and creating governmental regulations. Strict government intervention has worked in some settings with certain issues, but widespread problems require broader solutions. A group of social scientists known as environmental psychologists propose that policymakers must integrate multiple frameworks in order to create solution strategies. Most environmental issues are more about changing human behavior than saving the planet. Environmental psychologists suggest that Garrett Hardin's tragedy of the commons demonstrated the futility of trying to manage resources when every individual stands to gain through self interest. Hardin believed the only solution was to provide incentives for restraint and imagined that the government would have to assume that role. He was not optimistic about the chances of solving the tragedy of the commons in specific instances or in general. Hardin suggested that only by stopping human population growth could a global tragedy of the commons be averted. Slowing population growth might ease the situation, but it cannot happen overnight, and even that step requires change in human behavior. Social scientists use the tragedy of the commons as a cautionary tale about making the same mistake Hardin made in viewing the range of possible solutions too narrowly.

In order to make changes in how people act, environmental psychologists have begun to propose a combined solutions approach. Through extensive research, they have identified four components that lie at the heart of human behaviors as those behaviors relate to the environment. No single solution by itself can alter the activities of a majority of people. Combining the solutions, however, provides an effective way of reaching a large population, connecting the problem with current behaviors, and offering alternatives that will improve conditions. The four solutions actually serve as *frameworks* within a larger solution strategy.

The solution employed most frequently in the United States since the late 1960s is government regulation. This approach to environmental problems has succeeded in improving concerns such as air and water quality and the release of toxins. This success results primarily from the identification of major polluters. Laws require them to improve disposal practices or face stiff penalties. Government intervention can also serve as an incentive to companies that clean up voluntarily by rewarding them with tax credits and government contracts for their products. In this way, regulators have tools that include both positive and negative incentives. When the government can identify *pollution* sources easily and act quickly through legislation, this approach effectively addresses environmental problems. However, pollution sources are not always easy to identify and legislation is sometimes difficult to enact. In addition, enforcement of new legislation typically requires government agencies to develop specific rules before regulations can take shape. These steps become bogged down all too easily in political bargaining. Finally, the legal system may reveal loopholes as cases reach the courts. As Hardin noted, governmental regulation will not solve all environmental problems. Yet, it remains an important framework within larger solution strategies.

A second solution that has gained enormous popularity in recent decades is environmental education. Information on how environmental problems arise and the risks they bring plays an essential role in raising people's awareness. Health concerns related to pollution in the air and water can be addressed through education. Informing people of natural habitat loss that may affect recreation opportunities or community aesthetics raises awareness. Environmental educators have introduced topics like recycling and deforestation to the curriculum for young children. Lessons for children often connect a wonder of nature with a desire to be involved and do good deeds. A generation of young people have now grown up familiar with a wide range of environmental concerns and their causes. Awareness, however, only occasionally leads to action. Instead, sometimes education contributes to feelings of being overwhelmed by problems that have no clear solution, and the result is apathy rather than action. Critics of environmental education for children also suggest that such information is mere propaganda aimed at gullible youngsters. Social scientists have found that linking information directly to steps that will improve a situation tends to be more effective. For example, homeowners who see data on energy lost from poorly insulated homes may take steps to improve the insulation in their own homes if tips on how to seal doors and windows accompanies that data. In addition to putting information in the hands of people who can take action, this method also connects education to other incentives, such as saving money on energy costs.

A third solution that receives regular attention but often appears in more subtle ways is to appeal to people's values. Social scientists have examined a wide range of religious traditions searching for those that might be more responsive to environmental concerns. There have also been attempts to confirm the widespread belief that certain cultures hold environmental values dear while other cultures are antienvironmental. The most common examples of these are Native American cultures as embracing nature and Judeo-Christian cultures as dominating or destroying nature. Neither of these caricatures is accurate, and according to social scientists, most religious and cultural traditions contain a core message of stewardship of the earth. This value of caring for the environment may focus on agricultural traditions or passing viable resources on to future generations. Whatever the case, making values a substantial framework for a solution strategy requires helping people identify and articulate those ideals as they relate to stewardship, efficiency, or conservation. Simply telling people from a Judeo-Christian, Muslim, Buddhist, or other background to believe what American Indians believe about the earth would rarely succeed. Changing people's values is often more difficult than changing their behaviors. This is especially the case when certain values, such as conservation of resources, can appear to stand in conflict with other values, such as providing a comfortable modern home for one's family. Since most people already have values that align in some way with environmental concerns, incorporating those into a larger strategy provides a more promising outlook.

The fourth solution that appears regularly in connection with certain environmental problems is community management. Without direct government involvement, smaller groups of people can devise procedures for responding to local issues. They can also enforce those procedures more effectively than a larger governing body. Many of the most successful community management plans emerge between neighbors talking on the front steps of their homes or at a neighborhood picnic. The plans often address issues like safety and crime prevention, parking on the street, making space for children to play across several backyards, or minimizing noise at night. The issues could also be environmental in nature: preserving a plot of prairie or woods from development, minimiz-

Cape buffalo graze in the shadow of Mt. Kilimanjaro. The mountain, located in Tanzania near the Kenyan border, is the highest in Africa. Ancient ice near the summit may be gone in 15 years at the present rate of melting. (Corel)

ing the use of toxic fertilizers and weed killers on lawns where children play, or installing more energy efficient appliances in homes to reduce the community's electricity needs and costs. Because communities vary greatly in structure, effective management on this level rarely emerges on its own. Especially in the United States, where suburbanization coupled with privacy concerns, career focus, and smaller nuclear families have tended to isolate people from their neighbors, community management is unlikely to encourage change without some larger stimulus.

The most reasonable solution strategy combines these four separate approaches as frameworks to a larger solution. Just as each approach faces limitations, each also offers a means of overcoming the limitations of the others. Where government intervention can seem impersonal and oppressive, community management can offer more local insights. Where education sometimes appears partial to one ideology, the values approach introduces common principles of belief. Social scientists point to success stories in changing behavior as examples where a combination of approaches contributed to effective outcomes. Most curbside recycling programs now mimic the early projects that incorporated these approaches. Local governments provided incentives in the form of reduced waste disposal fees to households that would separate recyclable materials from their municipally disposed garbage. Information on the benefits of recycling demonstrated how the program would be helpful in reducing the need for new landfills, but also how recycling could start to affect how raw materials were acquired. Recycling paper and steel products would minimize cutting forests and mining in remote areas. Aluminum recycling could dramatically reduce the electricity needs of refining aluminum from raw mineral sources. Each of these benefits was in turn connected to values of conservation, efficiency, and even wilderness preservation. Another key component of these programs involved the distribution of highly visible sorting containers to facilitate separating recyclables from garbage and signs to advertise curbside pickup in local neighborhoods. Families could feel a sense of pride in reminding their neighbors of the next recycling day and having their designated recycling container on the curb first thing in the morning. The success of those programs is perhaps best illustrated by the fact that these explicit practices are becoming less visible as recycling has become a familiar part of everyday life for millions of people. The behavior of throwing the morning paper or an aluminum can in the trash is almost completely transformed for people in most areas. Many people are, in fact, reluctant to throw any recyclable item into the garbage, even where no recycling bins are available.

Conclusion

McKibben's *End of Nature* appeared in a new edition in 1999. His argument that human activities had unmade that natural world was bolstered by a decade of soaring temperatures and erratic weather patterns. These signs of global warming received multiple interpretations. Like McKibben, the Sierra Club maintained a stance that the dire predictions of the late 1980s were already coming true. By 2003, the Sierra Club highlighted melting glaciers in the mountains of Montana and the Cascade Range across California, Oregon, and Washington. In Europe, a 5,300-year-old human body was discovered in melting ice a decade earlier. Ice atop Africa's Mount Kilimanjaro has receded over 80 percent and, at the current rate, will be gone completely in another 15 years. Examples in South America include the melting glaciers of the Andes Mountains that supply water to much of Bolivia and Peru. That water source may be gone forever if ice continues to melt, and in the meantime, flooding also creates cause for concern. In the Arctic, changes in the distribution of ice threaten wildlife, and massive ice sheets are breaking off from Antarctica at unprecedented rates.

All of this adds up to reasonable cause for concern. The hope that environmental scientists might provide conclusive proof that global warming is hap-

This 5,000-year-old man, dubbed Otzi, was discovered on the Similaun Glacier in northern Italy by two hikers. Various items were found with the mummified man including a wooden backpack, copper ax, a dagger, and a bow with arrows. The emergence of these remains from a millennia old ice field suggest that the earth is warming. (Corbis)

pening, however, tends to miss the point of what science can offer in this debate. Since people might describe virtually any scientific idea as "just a theory," the burden of proof when massive economic and social structures are at stake cannot rest wholly on scientists. Instead, policymakers and the general public must begin to accept the evidence of science as a meaningful indicator of what skeptical specialists think is happening. Since scientists have made mistakes in the past, and since they are human, we can be reasonably certain that they will make mistakes in the present and the future. When the evidence from a wide range of specialized fields begins to coincide upon a common understanding of a situation, however, the public might be wise to direct its own skepticism toward critics of such generally accepted scientific conclusions.

Bibliographic Essay

A useful overview of atmospheric science and climate is S. George Philander, *Is the Temperature Rising?: The Uncertain Science of Global Warming* (Princeton: Princeton University, 1998). Another overview providing a more European perspective is Frances Drake, *Global Warming: The Science of Climate Change* (New York: Oxford University, 2000). A recent book that follows the rising awareness of global warming like a mystery novel is Spencer R. Weart, *The Discovery of Global Warming* (Cambridge, Massachusetts: Harvard University, 2003). James Rodger Fleming has written extensively on the history of climatology, with a book focusing specifically on climate change, *Historical Perspectives on Climate Change* (New York: Oxford University, 1998).

Details of the scientific findings of the Intergovernmental Panel on Climate Change appear in their three reports: *Climate Change: The IPCC Scientific Assessment*, ed. J. T. Houghton, G. J. Jenkins, and J. J. Ephraums (New York: Cambridge University, 1990); *Climate Change 1995: The Science of Climate Change*, ed. J. T. Houghton, et al. (New York: Cambridge University, 1996); and *Climate Change 2001: Synthesis Report*, ed. Robert T. Watson, et al. (New York: Cambridge University, 2001). Each of those reports contains extensive bibliographies of specific scientific studies of atmospheric conditions. These reports and more are available online at http://www.ipcc.ch/.

Climate change is a regular topic in a variety of environmental news media, such as the Sierra Club's magazine, *Sierra*, and publications from the Nature Conservancy.

Epilogue

This book has examined a range of issues that provide a glimpse of the intersection between environmental science and society. Topics like acid rain, biodiversity, Gaia, and climate change provide a sense of what is known. Such topics also hint at what is at stake and what the costs might be of finding alternatives to current practices or solutions to current problems. At the same time, it is important to point out that these examples merely represent a larger set of practices and problems that pose challenges to scientists and to society. This sample identifies some components of environmental science that can be consistently brought to bear on a wider range of issues. The examples discussed here also demonstrate how and to what extent social responses remain significant to environmental issues.

Development of historical awareness may provide a basis for deciding what social responses are appropriate in a context of scientific uncertainty or political controversy. Scientists and citizens cannot always plan for the difficulties that arise when they uncover and begin to debate environmental problems. Since these problems can lead society on different trajectories, solution plans that are intended to correct one and all will often fail. Instead, if history has lessons to teach, flexibility will provide a more fruitful starting point than rigid policies, regulations, and frameworks.

Emerging concerns may yet submit to the existing frameworks of how science and society deal with environmental problems. More likely, such issues will expand the challenge to scientists, governmental leadership, and nongovernmental organizations alike. To take just one example, the development of genetically modified organisms, the need for even greater interaction between science and social leadership quickly becomes evident. Genetically modified organisms result from manipulation of genetic information that transforms an existing species, typically by inserting genes for characteristics previously unknown to the original species. So far, crops that are resistant to certain pests or that tolerate certain chemical herbicides have created the promise of higher yields. Since human agriculture has produced hybrids for millennia, the emergence of altered crops carries a certain sense of the status quo. However, the products of genetic manipulation are not hybrids in the traditional sense, for they bring together characteristics of organisms that could never interact reproductively in nature. Certain extreme examples highlight this incongruence, such as rabbits that glow in the dark as a result of being injected with a gene for phosphorescence from a jellyfish.

Other examples of genetic engineering seem less bizarre, and their resulting characteristics offer great hope, such as rice that produces excess vitamin A, a nutrient typically deficient in undernourished people. Working to solve world hunger and malnourishment stands as a laudable goal for science. Here is where science and social needs require more intense scrutiny. Some reports of the vitamin A-enhanced rice claimed that it would save children from blindness. Calculations of the amount of this rice needed for a child soon revealed that such a claim overstated the contribution enhanced rice could make in a child's diet. Responses to those calculations attempted to clarify that the scientists responsible for developing this rice did not claim to solve world hunger and save children from blindness, but rather that they hoped the enhanced rice could supplement diets as a part of ongoing efforts to provide a more varied supply of nutritional foods. On one hand, corporations behind the development of genetically modified organisms hoped to recover research costs by promising revolutionary products. On the other hand, skeptics reacted to those promises by questioning the apparent greed of corporations. Some skeptics dubbed the products as "frankenfoods," creating an image of science gone mad. Environmental groups and corporate sponsors of the research remained polarized, obscuring the potential for more constrained testing of certain products within carefully monitored settings.

While the issue of genetically modified organisms departs from the more familiar range of topics that environmental scientists and environmentalists have pursued, at its core is the question of what constitutes a natural entity. Genetically modified organisms resist easy categorization. Whatever definition scientists or critics choose to impose, a range of new organisms might fall across a spectrum of natural species and synthetic products.

At the same time, even traditional natural entities have become difficult to define at the intersection of science, public perception, and government regulation. The idea of wilderness serves as a prime example. Historians, sociologists, geographers, and philosophers have formalized a line of inquiry that continues to inspire debate and dissension. Strange as it may seem at first glance, some argue that wilderness has become an icon for what humans construct out of their expectations for nature. Forests, prairies, tundra, and deserts uninhabited by humans may be real places, but currently, few are separated from all human interference. As a result, some people argue that these places no longer qualify as wilderness; wilderness may have ceased to exist. More radically, according to some people, wilderness never existed. Wilderness, as a human construct, has a history of its own. That history is an account of how people have perceived the places where they do not live as different from the places where they build communities and cultures by reshaping or working within the constraints of the landscape. Here again, wilderness is just an example of the broader issue of whether nature is really "out there."

Human efforts to preserve, protect, conserve, and restore nature during the past century or more produced greater knowledge of nature's assets, even if those efforts did not always achieve the goals set by scientists, government officials, and environmentalists. Preservation and protecting nature often took the form of setting aside tracts of land where particularly valuable species might go on living undisturbed by human influence. The lack of disturbance by humans, however, did not consistently translate into the preservation of the species. Unforeseen factors disrupt natural systems, even in the absence of direct human contact. Those disruptions, sometimes caused by other species and other times caused by such unpredictable forces as the weather, have led to significant rethinking of the stability that so many environmental planners hope to achieve.

According to most ecologists today, stability in nature is an illusion of locality and time scale. When attempting to preserve a species, environmental planners may need to focus on the many alternative situations that arise in nature rather than attempt to stabilize a given area. In places where human disturbances, or even preservation efforts, have changed conditions for threatened species, attempts to restore habitat become more complicated. Ecologist Daniel Botkin suggests that nature is dynamic, and restoration requires change. Kirtland's warbler, an endangered species of songbird that nests in the jack pine forests of Michigan, depends upon periodic fires to open up the forest. Since fire disturbs the habitat, people suppressed fires for almost one hundred years for the expressed purpose of protecting the warbler habitat. During that time, warbler numbers actually declined. In 1951, a count revealed only 400 singing males. A similar count in 1971 found just over 200 males. The numbers continued to decline in the decades that followed. Only recently, according to Botkin, have ecologists recognized the importance of limited regular burning of the forest to produce attractive nesting sites for the warbler population.

Similarly, ecologists now challenge the assumption that salmon always return to their native stream. When the fish mistakenly take the wrong fork on their return trip upstream, they sometimes find themselves in more favorable waters. Since landslides, sedimentation, and changes in water level can disrupt salmon habitat, species have never had a guarantee of success in finding their native stream. Just as likely, changes in conditions could make previously unsuitable waters into an ideal breeding stream. Salmon evolved with a certain level of flexibility in discovering new breeding waters from generation to generation, even if most individuals did return to their native streams. Since change and habitat disturbance can benefit some species on occasion, the argument that human-induced change can be beneficial might logically follow. Actual examples of beneficial change in nature become rare, however, as the source of those changes increasingly involves human activities that rapidly and radically alter conditions that previously persisted for centuries.

Considering change in natural systems leads ecologists to think historically. Donald Worster, an environmental historian, has argued that the very act of thinking about nature historically has created dilemmas in how humans approach problems of preserving or restoring natural places. Connecting past landscapes with a wide range of possible futures becomes a controversial enterprise. While ecologists recognize that the present-day natural world is the way it is largely due to its history, they must increasingly acknowledge that the natural world is at least partly constructed by human activities and perceptions. Ecologists who think historically have influenced historians of human culture who think ecologically. Worster's point is that historians tend to focus on the changes that have taken place: changes in nature itself, changes in human activities, and changes in human perceptions of nature. He speculates that historical studies that focused instead on stability and stasis might uncover an equally large, if less compelling, story. With change as the focus, however, both ecologists and historians conclude that change has consequences. The strength of this conclusion has implications for society through public perception and response to crises as well as through policymaking.

The persistent belief that nature is "out there" and that scientists can study nature objectively is now brought into question. If nature is as much a construct of human perception as an external reality, it is subject to the theory of historical relativism that has figured prominently in postmodern scholarship. Worster provides a vivid example of what this might mean for efforts to preserve and even to study nature. He writes, "Disneyland, by the theory of historical relativism, is as legitimate as Yellowstone National Park, a wheat field is as legitimate as a prairie, a megalopolis of thirty million people is as legitimate as a village" ("Nature," 78). Considering these different landscapes equally legitimate, however, does not make them equally desirable or sustainable. Worster goes on to argue that ecology and history both reveal the importance of interdependency: of cultures, of species, of communities. As a consequence, any criteria for deciding how to preserve remnants of the natural world must depend on how interdependent parts might still function. Preserving or recreating interdependently functioning systems will involve a cost. That cost can be compared to the cost of letting the interdependence disintegrate entirely and replacing it with some artificial system.

Using the criteria of interdependence also suggests an important role for history. In the past, the natural world persisted for eons. While that persistence was not always in a steady state, certain systems were long-lived, providing models for the present and future. Just as the wings of birds provide a model for people hoping to design airplanes, tallgrass prairies can provide a model for human societies hoping to limit erosion and maintain adequate moisture levels in the soil. It would be too costly to restore all the tallgrass prairies of the Great Plains, but it has already proven extremely costly to try and maintain those areas solely for agricultural use. Using as a guide the history and ecology of the region, environmental science can assist society in developing a sustainable common ground.

Aldo Leopold established a baseline for expectations of science and society near the middle of the twentieth century. Referring to the differences in those expectations, he wrote, "The ordinary citizen today assumes that science knows what makes the community clock tick; the scientist is equally sure that he does not. He knows that the biotic mechanism is so complex that its workings may never be fully understood" (*Sand County Almanac*, 241). That statement reflects the confidence of the public in science, and scientists' own awareness of the limitations of their knowledge. Since the 1960s, the public has grown more wary of science and its promises to control nature. From *Silent Spring* to global warming, society has felt the effects of unfulfilled promises and the uncertainty of long-term predictions. Scientists, meanwhile, have focused on more specialized subfields and, as a result, have developed more confidence in their understanding of these confined areas. Perhaps, however, Leopold's observation still reflects our fondest wishes. Despite evidence to the contrary, people continue to hope that science will provide the answers. Certainly, scientists continue to confront complexity in the natural world, and they have found the social and political world at least equally perplexing.

Bibliographic Essay

A large amount of literature on genetically modified foods has recently emerged as part of the current environmental movement. A recent assessment by two scientists is offered by Nina V. Fedoroff and Nancy Marie Brown, *Mendel in the Kitchen: A Scientist's View of Genetically Modified Foods* (Washington, D.C.: Joseph Henry, 2004). The Institute of Medicine (U.S.) Committee on Identifying and Assessing Unintended Effects of Genetically Engineered Foods on Human Health has recently published a study entitled *Safety of Genetically Engineered Foods: Approaches to Assessing Unintended Health Effects* (Washington, D.C.: National Academies, 2004); and an examination of some of the ethical, economic, and political issues comes from Rahul K. Dhanda, *Guiding Icarus: Merging*

Bioethics with Corporate Interests (New York: John Wiley & Sons, 2002). For more focus on the production of these foods, see a volume edited by Gerald C. Nelson, *Genetically Modified Organisms in Agriculture: Economics and Politics* (San Diego: Academic, 2001).

Further discussion of how history can inform environmental science is provided by Donald Worster, "Nature and the Disorder of History," in *Reinventing Nature? Responses to Postmodern Deconstruction*, edited by Michael Soulé and Gary Lease (Washington, D.C.: Island, 1995), 65–85. Other essays in the Soulé and Lease volume discuss the significance of scientific evidence in light of criticism from scholars outside environmental science. Aldo Leopold's *A Sand County Almanac* (New York: Oxford, 1949) remains a classic.

Glossary

abiotic factors   components of the environment that are not living but contribute significantly to the cycling of nutrients and flow of energy in an ecosystem.

acid neutralizing capacity   the ability for soils to act as a buffer against acid rain.

acid rain   rain and other forms of precipitation with a higher than average acidic content; in most cases a pH below 5.6.

aquatic   having to do with fresh water, as communities in lakes, streams, and rivers.

Atomic Age   the period beginning with the first nuclear test in New Mexico; its significance is usually described as the period when humans learned that they could "destroy the Earth," or at least make the planet unlivable.

balance of nature   generally used to describe a state in which nature is untouched by human interference, where population numbers remain essentially stable; later, ecologists acknowledged that the balance was dynamic, regulated by competition between species, food supplies, and climate; meanings and implications of this phrase vary enormously.

bioaccumulation   see *biological magnification*.

biodiversity   the number of different species inhabiting an area; once considered simply as "diversity" or "species richness"; the term has become closely connected to preservation efforts that link the abundance of different types of living organisms to a value that makes an area worth protecting.

bio-ecology   a term used to describe plant-animal studies in ecology, beginning with the work of Victor Shelford in the 1930s.

biogeography   the study of the distribution of plants and animals across diverse climatic and geographical zones.

biological magnification   the process whereby minute amounts of a substance, often a radioactive isotope of a common element or a toxic chemical, move through a food web reaching higher concentrations at higher trophic levels due to the inability of organisms to remove the substance from tissues once it begins to accumulate.

biology   the study of living things, meant in the early nineteenth century to refer especially to studies that would involve both plants and animals.

biomass   a measure of the living material in an ecosystem, usually made by measuring a sample and estimating the percentage of the total represented in that sample.

biosphere   the unit of biological organization that contains all living organisms, ranging from bacteria deep within the Earth's crust or in vents at the bottom of the ocean to the highest levels of the atmosphere.

biotic community   all the plants and animals living in an area; the term emphasizes their interdependent relationships.

botany   the study of plants; a specialty of natural history that dates back at least to Aristotle.

carrying capacity   the number of organisms a given environment could maintain; in many situations where this term was initially applied, the concern was for profitable maintenance of livestock and sometimes of game.

catastrophism   an explanation for the current state of nature based upon past periods of worldwide episodes of flooding and geological forces; those episodes would not be matched by any currently observed phenomena.

causation   two phenomena directly linked as cause and effect by evidence of the undisputed influence of one upon the other.

chlorofluorocarbons   synthetic compounds considered nonreactive and useful in a variety of applications as coolants and propellants, and in the production of styrofoam; shown to react with ozone at high altitudes.

classification   the process of naming and grouping plants, animals, and minerals, which dates back to antiquity but was standardized by Linnaeus.

climate   conditions of moisture and temperature that persist over large regions of the planet for relatively long periods of time.

climax stage   the formation of plants assumed to represent the most stable and lasting assemblage in a given area based on climatic conditions.

Cold War   beginning at the close of World War II, the period of limited political relations between communist countries (primarily the Soviet Union) and the countries of the so-called West (particularly western Europe and the United States).

community   in ecology, a study of plants and animals that takes interdependent relationships into account, as contrasted with studies of individual species or of plants and animals considered separately in botany or zoology.

conservation   a movement to protect natural resources largely motivated by the utility and value of those resources for the present and future generations.

consumers   in ecology, the organisms that eat other organisms; those that eat mostly plants and other producers are first-level consumers; those that eat other consumers are second- or third-level consumers.

correlation   two phenomena related by their shared response to additional influences, potentially independent of one another.

DDT   a chemical, dichloro-diphenyl-trichloro-ethane, found to be effective in destroying insect populations, later proven to accumulate in the tissues of species higher up in the food chain, causing death to songbirds and reproductive failure of predatory birds.

decomposers   organisms that live on dead organic material, converting nutrients back into compounds that enrich soils and provide the building blocks for other organisms.

diversity   the range of plants and animals, including the variety of forms within each species as well as the variety of different species.

dose-response relationships   a comparison of the effects of different levels of radiation (or other forms of treatments) on an organism's ability to function properly.

ecology   the study of the interactions between organisms of different species and their environments; a self-conscious science beginning in the 1890s primarily focused on plants; animal ecology developed in the 1910s and 1920s; other definitions include: new natural history; the study of whole organisms in the field; and the connections among all living things.

economy   the interactions of systems, as monetary systems, where checks and balances are regulated by production and consumption; used as an analogy to illustrate various aspects of natural history.

ecosphere   a range of habitats circling the planet, connecting all living communities and ecosystems across oceans and continents as a global system.

ecosystem   the habitat and community as an interacting unit, especially considering the cycling of nutrients (nitrogen, carbon, oxygen, etc.) and the flow of energy from producers to consumers (see *food chain*).

entomology   the study of insects; this science became increasingly focused on the control of pest species and was often referred to as "economic entomology" in the early twentieth century.

environmentalism   a social movement focusing on concerns about changes in the natural world as a result of human activities; the movement gained momentum in the 1960s as scientists revealed a series of destructive aspects of human action.

eukaryotes   organisms made up of cells that contain a full complement of organelles, with chromosomes inside a nuclear membrane that bounds the nucleus of each cell.

eutrophication   a process of collapse in aquatic ecosystems, often caused initially by an increase in nutrients that leads to rapid algae growth and ultimately to oxygen depletion and fish kill as the algae dies and decays.

evolution   changes in species over time resulting from one of several proposed mechanisms; Darwin's proposed mechanism for evolution was natural selection.

fauna   plant life as studied by botanists and plant ecologists.

feedback   processes within a system that result from the system's functions and, in turn, cause the system to function with a changed level of efficiency; positive feedback increases the output of the system, which would increase the level of feedback; negative feedback decreases the output of the system, thus decreasing the level of feedback.

flora   animal life as studied by zoologists and animal ecologists.

floristic phytogeography   the study of how flowering plants are distributed geographically.

food chain   a series of relationships between organisms designated by what they eat and by what eats them; commonly expressed by linking *producers* to *consumers* of increasingly higher orders; multiple food chains generally intersect to form a food web.

food web   the more complex matrix of producers and consumers, showing how food chains interact in communities.

formation   a community of multiple populations of species; in the work of Frederic Clements, this is usually plants.

framework   an approach to problem solving based in a particular disciplinary context, with specific tools or methods consistently applied.

Gaia hypothesis   a description of the components of the biosphere operating as an individual organism, with features that maintain conditions at a steady state.

geology   the study of the earth's surface; a specialty of natural history that took on great significance in the late eighteenth century and early nineteenth century as the age of the earth became a topic of interest to naturalists.

global climate change   the phrase adopted by most scientists to indicate the alterations in the Earth's atmosphere believed by most scientists to be the result of human activities that increase concentrations of carbon dioxide, methane, and other gases that prevent heat from radiating back into space; see also *global warming*.

global warming   a widely used expression for atmospheric changes with a trend toward general increases in temperature.

gradualism   an explanation that involves slow and relatively steady change in natural phenomena.

greenhouse effect   a popular expression used to describe the radiation of heat from the Earth that is trapped by atmospheric concentrations of gases such as carbon dioxide, methane, and water vapor.

herbicide a chemical agent intended to kill plants, particularly those considered to be weeds that interfere with domestic crop productivity.

holistic/holism a philosophy of nature in which all parts are vitally connected and cannot survive without the contributions of the others, as in a single organism.

infrared radiation with longer wavelengths than visible light, this radiation is detected as the heat given off by objects exposed to energy sources such as sunlight.

irruption a population cycle where a rapid increase, caused by removal of enemies or unanticipated excess of growth resources, is followed by an equally rapid decline when resources are depleted.

island biogeography the study of species diversity on land units that are separated from other land units by geographic features (such as but not limited to open ocean) that limit migration; by obtaining an estimate of certain features, such as the initial number of species on the island, migration rates, and the rate of evolutionary change, the effect of those features on the species composition of a given area can be predicted.

isotopes forms of an element that contain varying numbers of neutrons; certain forms decay spontaneously to give off radiation.

land ethic a principle introduced by Aldo Leopold in the 1940s, suggesting that long-term protection of *biotic communities* should supercede short-term human needs.

life cycle a description of the different stages of the life of an organism, often emphasizing different biological and ecological needs during each stage.

logarithmic a comparative scale, such as the *pH* scale, where increments of one on the scale represent a tenfold change in the corresponding concentration.

materialism a philosophical commitment to explanations that result from phenomena that can be sensed and explained within the natural world without recourse to supernatural explanations.

meteorology the study of the atmosphere, so-named because the realm of the sky where weather develops is also the lower layers where meteors glow brightly when they enter the atmosphere.

microcosm a smaller version, where analogies can point to the complexities and intricacies of a larger system, but where certain components can be studied in detail with greater ease.

natural history the study of animals, plants, and minerals primarily through observation; the primary activities of natural history were collection and description.

natural philosophy descriptions of nature that reflected on causes and searched for underlying principles; natural philosophers eventually adopted mathematics as a part of these descriptions.

natural selection Darwin's mechanism for evolution; individuals of a species vary, giving a survival advantage to certain individuals, and those individuals will be more likely to survive and reproduce, passing those advantages on to their offspring.

natural theology an approach to the study of nature based upon the search for God's benevolence in creation; explanations for natural phenomena generally derive from their usefulness within a supernatural plan.

naturalists professional and amateur practitioners of natural history studies, especially prior to the twentieth century.

nature the varied components of the universe generally regarded as separate from but affected increasingly by human activity.

niche the functional position of a species within a community, where competition for resources is minimized by specializations of the species.

organicism a philosophical viewpoint in science that suggests entities without an individual identity can be considered organisms for the sake of useful analogies; in a stronger version, these entities (such as a community of plants, animals, and other living beings) are collectively considered to be organisms in actual practice.

ornithology the study of birds, a specialty of natural history that gained prominence in the nineteenth century and led some scientists to actively demand legislation protecting birds in the United States by the end of the century.

ozone a molecule made up of three oxygen atoms (O_3); at high altitudes, provides a shield against ultraviolet radiation from the sun; near the ground, contributes to smog and haze, creating difficulty in breathing for people with respiratory ailments.

parts per million units of measurement used for very small amounts of a substance in a given solution; one part per million is equivalent to 0.0001 percent.

pesticide a chemical agent intended to kill pests, particularly insects.

pH a scale for comparing the acidity or alkalinity of a solution ranging from zero (acidic) to 14 (basic), based upon the concentration of hydrogen ions (H+) in the solution.

photosynthesis the process of converting carbon dioxide and water into organic sugars using energy derived from light, giving off oxygen as a waste product; multicellular and unicellular plants as well as cyanobacteria undergo photosynthesis.

pioneer plants the first plant species to thrive in a barren area following a disruption that removes plant formations from later successional stages.

pollution   products of human activities that alter the environment in ways generally accepted as negative to human health.

predator   a meat-eating animal, a carnivore; generally paired with the animals it eats (prey), as in predator-prey relations.

preservation   in the early twentieth century, a movement to protect natural wonders that conflicted with conservation, where conservation emphasized the utilitarian value of natural resources.

producers   plants and some microorganisms that convert inorganic molecules to energy-storing organic molecules; serve as food for consumers.

professionalization   the process of establishing a community of scientists with consistent standards for education and employment.

progressivism   a social reform movement of the early twentieth century, when advocates believed that scientific management and efficiency held the key to improvement in society.

pyramid of numbers   a comparison among species at different levels of a *food chain*, typically with the greatest number among the producers and considerably fewer at higher levels of consumers.

quadrat   a small unit of an ecological *community*, usually one meter square, that could be studied intimately and quantitatively in order to describe the larger community.

quantitative research   studies of the natural world that emphasize measurement and counting in the collection of data, as opposed to qualitative research, where observations are not reduced to numerical data.

radioactivity   energy in various forms caused by the decay of unstable isotopes that can disrupt normal physiological processes.

radiotracers   radioactive atoms that otherwise function in biological molecules and can be used to track the movement of nutrients in an ecosystem.

reductionist approach/reductionism   an approach to the natural world that depends upon scientists' ability to study small components of systems with the assumption that the parts will add up to explain the whole.

science   the practice of studying nature, which has changed dramatically over time from philosophical and descriptive practices to experimental and mathematical approaches.

scientific models   simplifications of natural phenomena that can be used to represent and forecast conditions that are otherwise too large, too small, occur too rapidly, or occur too slowly to be observed directly.

scientist   a person who studies the natural world with certain commitments to explanations that satisfy a skeptical community; the commitments of scientists have changed over time but have generally included logic, repeatability, and some level of *materialism* and mechanism since the seventeenth century.

succession   a more or less deterministic process by which communities originate, develop, and reach a stable equilibrium, dependent upon climatic conditions ranging from soil type and precipitation to disturbances of both human and natural origin.

systems ecology   a highly quantitative approach to ecosystems, where nutrients and energy are modeled mathematically.

taxonomy   the study of the relations of organisms that assists in classification; for evolutionary biologists, relations represent the products of descent from a common ancestor.

temperate   climates with moderate levels of precipitation, where temperatures range from cold in the winter to warm in the summer, generally in latitudes between 30 and 60 degrees north and south of the equator.

tragedy of the commons   a situation arising out of the shared availability of a given resource, where all users benefit from taking more than their individual share of the resource but all eventually suffer when the resource is exhausted by overuse.

trophic-dynamic   an approach to ecosystem studies that emphasized the changing nature of energy transfers between organisms at different levels within the system.

tropical   climates where temperatures remain warm year round, generally within 30 degrees north and south of the equator.

ultraviolet radiation   high energy light waves that can disrupt biological processes; most of these waves are blocked by the earth's atmospheric ozone layer.

uniformitarianism   an explanation that assumes that phenomena observed at present will continue at similar rates and are responsible for all past changes in nature.

visible light   the form of radiation from the sun that reaches the Earth as white light, containing the colors of the rainbow.

weather   local conditions of moisture and temperature that can change relatively rapidly as a result of short-term influences.

wildlife management   a field of science made up primarily of professional zoologists and ecologists who focus their research on practical issues related to matching animal populations with habitat and human demands.

zoology   the study of animals, based in pre-twentieth century *natural history*.

Documents

Conrad Gesner's encyclopedic descriptions of plants and animals reflects the classical form of natural history, dating back to the Ancient Greeks. He included familiar species and well-known facts. In addition, Gesner often described animal and plant forms that did not exist in nature but were suspected to exist due to their frequent appearance in histories, both factual and fanciful. Examples appear in this excerpt. Gesner's inclusiveness, placing werewolves and unicorns alongside bears, wolves, and zebras, was considered an appropriate means of chronicling natural history during a period of exploration and uncertainty.

The History of Four-Footed Beasts and Serpents and Insects: Taken Principally from the historiæ animalium *of Conrad Gesner* (c. 1551)

CONRAD GESNER

Of the Elephant

There is no creature among all the Beasts of the world, which hath so great and ample demonstration of the power and wisdom of Almighty God as the Elephant: both for proportion of body and disposition of spirit; and it is admirable to behold the industry of our ancient fore-fathers, and noble desire to benefit us their posterity, by searching into the qualities of every Beast, to discover what benefits or harms may come by them to mankinde; having never been afraid either of the wildest, but they tamed them; the fiercest, but they ruled them; and the greatest, but they also set upon them. Witness for this part the Elephant, being like a living Mountain in quantity and outward appearance, yet by them so handled, as no little Dog became more serviceable and tractable.

Among all the *Europœans* the first possessor of Elephants, was *Alexander Magnus*, and after him *Antigonus*, and before the *Macedonians* came into *Asia*, no people of the world, except the *Africans* and the *Indians*, had ever seen Elephants. When *Fabritius* was sent by the *Romans* to *King Pyrrhus* in Ambassage, *Pyrrhus* offered to him a great sum of money, to prevent the War, but he refused private gain; and preferred the service of his Countrey: the next day he brought him into his presence, and thinking to terrifie him, placed behinde him a great Elephant, shadowed with cloth of Arras; the cloth was drawn, and the huge Beast instantly laid his trunk upon the head of *Fabritius*, sending forth a terrible and direful voyce: whereas *Fabritius* laughing perceiving the policy of the King, gently made this speech:

Neque beri aurum, neque hodie bestia me emovit.

I was neither tempted with thy Gold yesterday, nor terrified with the sight of this Beast today: and so afterward *Pyrrhus* was overcome in War by the *Romans*, and *Manlius Curius Dentatus* did first of all bring Elephants in Triumph to *Rome*, calling them *Lucanæ Boves*, Oxen of the Wood, about the 472. year of the City; and afterward in the year of *Romes* building 502. When *Metellus* was high Priest, and overthrew the *Carthaginians* in *Sicily*, there were 142 Elephants brought in Ships to *Rome* and led in triumph, which *Lucius Piso* afterward, to take away from the people opinions of the fear of them, caused them to be brought to the stage to open view and handling, and so slain; which thing *Pompey* did also by the slaughter of five hundred Lions and Elephants together; so that in the time of *Gordianus*, it was no wonder to see thirty and two of them at one time.

An Elephant is by the *Hebrews* called *Behemah*, by way of excellency, as the *Latins* for the same cause call him *Bellua*, the *Chaldeans* for the same word, Deut. 14. translate *Beira;* the *Arabians*, *Behitz;* the *Persians*, *Behad;* and the *Sepluagint*, *Ktene;* but the *Grecians* vulgarly *Elephas*, not *Quast Elebas*, because they joyn copulation in the water, but rather from the *Hebrew* word *Dephil;* signifying the Ivory tooth of an Elephant (as *Munster* well observeth). The *Hebrews* also use the word *Schen* for an Elephants tooth. Moreover *Hesychius* called an Elephant in the *Greek* tongue *Perissas;* the *Latins* do indifferently use *Elephas* and *Elephantus;* and it is said that *Elephantus* in the *Punick* tongue, signifieth *Cæsar:* whereupon when the Grandfather of *Julius Cæsar* had slain an Elephant, he had the name of *Cæsar* put upon him.

The *Italians* call this beast *Leofante*, or *Lionfante;* the *French*, *Elephante;* the *Germans*, *Helfant;* the *Illyrians*, *Slon*. We read but of three appellative names of Elephants; that is of one, called by *Alexander* the great *Ajax*, because he had read that the buckler of great *Ajax* was covered with an Elephants skin, about whose neck he put a Golden collar, and so sent him away with liberty. *Antiochus* one of *Alexanders* successors had two Elephants, one of them he likewise called *Ajax*, in imitation of *Alexander*; and the other *Patroclus*, of which two this story is reported by *Antipater*. That when *Antiochus* came to a certain ford or deep water, *Ajax* which was alway the Captain of the residue, having sounded the depth thereof, refused to passe over, and turned back again, then the King spake to the Elephants and pronounced, that he which would passe over should have principality over the residue: whereupon *Patroclus* gave the adventure, and passed over safely, and received from the King the silver trappings and all other prerogatives of principality; the other seeing it (which had always been chief till that time) preferred death before ignominy and disgrace, and so would never after eat meat but famished for sorrow.

They are bred in the hot Eastern Countries, for by reason they can endure no cold, they keep only in the East and South. Among all, the *Indian* Elephants are greatest, strongest, and tallest, and there are among them of two sorts, one greater (which are called *Prosii*) the other smaller (called *Taxilæ*). They be also bred in *Africa*, in *Lybia*, much greater then a *Nysæan* Horse, and yet every way inferiour to the *Indian;* for which cause, if an *African* Elephant do but see an *Indian*, he trembleth, and laboureth by all means to get out of his sight, as being guilty of their own weakness.

There are Elephants also in the Isle *Taprobane*, and in *Sumatra* in *Africa*. They are bred in *Lybia*, in *Æthiopia*, among the *Troglodytæ*, and in the Mountain *Atlas*, *Syrtes*, *Zames*, and *Sala*, the seven Mountains of *Tingitania*, and in the Countrey of *Basman*, subject to the great *Cham*. Some Authors affirm, that the *African* Elephants are much greater then the *Indian*, but with no greater reason then *Columelia* writeth, that there be as great beasts found in *Italy* as Elephants are: whereunto no sound Author ever yeelded.

Of all earthly creatures an Elephant is the greatest: for in *India* they are nine cubits high, and five cubits broad; in *Africa* fourteen or fifteen full spans, which is about eleven foot high and proportionable in breadth, which caused *Ælianus* to write, that one Elephant is as big as three Bugils; and among these the Males are ever greater then the Females. In the Kingdom of *Melinda* in *Africk*, there were two young ones not above six monthes old, whereof the least was as great as the greatest Ox, but his flesh was as much as you shall finde in two Oxen; the other was much grater.

Of the Wolf

A Wolf is called in *Hebrew Zeeb*, as it is said in *Gen. 49.* and among the *Chaldeans*, *Deeba* and *Deba*, among the *Arabians Dib*. The female is called *Zebah* a she-Wolf, and the masculine *Zeebim*, but in *Ezek. 22.* it is called *Zebeth*, that is to say, a Wolf. *Alsebha* (saith *And. Bellun.*) is a common name for all Four-footed beasts which do set on men, killing and tearing them in pieces, devouring them with their teeth and clawes, as a Lyon, a Wolf, a Tiger, and such like, whereon they are laid to have the behaviour of *Alsebha*, that is, wilde beasts which are fierce and cruel. From hence happily cometh it, that not only *Albertus*, but also some ignorant Writers do attribute unto a Wolf many things which *Aristotle* hath uttered concerning a Lyon.

Oppianus among the other kinde of Wolves hath demonstrated one which is bred in *Cilicia*. And also he doth write, that it is called in the mountains of *Taurus* and *Amanus*, *Chryseon*, that is to say, *Aureum*, but I conjecture that in those places it was called after the language of the *Hebrewes* or *Syrians* which do call *Sahab*, or *Sohab aurum*, and *Seeb Lupum* for a Wolf, or *Dabah*, or *Debab* for *Aurum:* They also do call *Deeb* or *Deeba* for a Wolf. *Dib* (othertherwise *Dijb*) is an *Arabian* or *Saracenican* word: Also the translation of this word in the book of medicines is divers, as *Adib, Adep, Adhip,* and *Aldip:* but I have preferred the last translation, which also *Bellunensis* doth use. *Aldip Alambat* doth signifie a mad or furious Wolf. The Wolf which *Oppianus* doth call *Aureum,* as I have said even now, doth seem to agree to this kinde, both by signification of the name *Aurum,* and also by the nature, because it doth go under a Dog close to the earth, to eschew the heat of the Summer, which *Oppianus* doth write, doth seek his food out of hollow places, as a Hyena or *Dabh* doth out of graves where the dead men are buryed. The golden coloured Wolf is also more rough and hairy then the residue, even as the Hyena is said to be rough and maned. And also these Wolves necks in *India* are maned, but it differeth according to the nation and colour where there are any Wolves at all.

    Lycos a Wolf among the *Grecians,* and *Lugos,* and *Lucania,* and *Lycos,* among some of the *Arabican* Writers, is borrowed from them, as *Munster* hath noted in his Lexicon of three languages. In *Italy* it is called *Lupo.* In *French Loup,* in *Spain Lobo,* in *Germany Vulff,* in *England Wolf.* In *Illyria Vulk,* as it were by a transposition of the letters of the *Greek* word. Now because both men, women, Cities, places, Mountains, Villages, and many artificiall instruments have their names from the *Latine* and *Greek* words of this beast, it is not vain or idle to touch both them and the derivation of them, before we proceed to the naturall story of this beast. *Lupus* as some say in *Latine* is *Quast Leopos,* Lyonfooted; because that it resembleth a Lyon in his feet, and therefore *Isidorus* writeth, that nothing liveth that it presseth or treadeth upon in wrath. Other derive it from *Lukes,* the light, because in the twilight of the evening or morning it devoureth his prey, avoiding both extreme light, as the noon day, and also extreme darknesse as the night. The *Grecians* do also call them *Nycterinoi canes,* dogs of the night. *Lupa* and *lupula* were the names of noble devouring Harlots, and from thenceforth cometh *Lupanar* for the stewes. It is doubtfull whether the nurse of *Romulus* and *Kemus* were a Harlot, or she-wolf, I rather think it was a Harlot then a Wolf that nursed those children. For we read of the wife of *Fostulus,* which was called *Laurentia,* after she had played the whore with certain Shepherds was called *Lupa.* In all Nations there are some mens names derived from Wolves, therefore we read of *Lupus* a *Roman* Poet, *Lupus Servatus* a Priest or Elder, of *Lupus de Oliveto* a *Spanish* Monk, of *Fulvus Lupinus* a *Roman,* and the *Germans* have *Vulf, Vulfe, Hart, Vulfegang.*

Source: Topsell, Edward. 1967. *The History of Four-footed Beasts. Taken principally from the* Historiæ Animalium *of Conrad Gesner. With a New Introduction by Willy Ley.* New York: Da Capo Press.

*John Ray contributed in important ways to the natural historical
studies of his generation. He carefully described observations of the
natural world and connected his belief that nature reflects the particular
intent and wisdom of the Creator. Ray's examples ranged widely from
plants to animals, as shown in the excerpt here, including the design of
the human form. He also emphasized the many human benefits of other
species, further demonstrating God's providence.*

The Wisdom of God Manifested in the Works of the Creation

JOHN RAY (1691)

To the fitness of all the parts and members of Animals to their respective uses
may also be referred another observation of the same *Aristotle*. All Animals
have even Feet, not more on one side than another; which if they had, would
either hinder their walking, or hang by not only useless, but also, burdensome.
For though a Creature might make limping shift to hop, suppose with three
Feet, yet nothing so conveniently or steddily to walk, or run, or indeed to
stand. So that we see, Nature hath made choice of what is most fit, proper and
useful. They have also not only an even number of Feet, answering by pairs one
to another, which is as well decent as convenient; but those too of an equal
length, I mean the several pairs; whereas were those on one side longer than
they on the other, it would have caused an inconvenient halting or limping in
their going.

I shall mention but one more Observation of *Aristotle*, that is, there is no
Creature only volatile, or no flying Animal but hath Feet as well as Wings, a
power of walking or creeping upon the Earth; because there is no food, or at
least not sufficient Food for them to be had always in the Air; or if in hot Countries we may suppose there is, the Air being never without store of Insects flying about in it, yet could such Birds take no Rest, for having no Feet, they could
not perch upon Trees, and if they should alight upon the Ground, they could by
no means raise themselves any more, as we see those Birds which have but
short Feet, as the *Swift* and *Martinet*, with difficulty do. Besides, they would
want means of breeding, having no where to lay their Eggs, to Sit, Hatch or
Brood their Young. As for the Story of the *Manucodiata* or *Bird* of *Paradise*,
which in the former Age was generally received and accepted for true, even by
the Learned, it is now discovered to be a Fable, and rejected and exploded by
all men: Those Birds being well known to have Legs and Feet, as well as others,
and those not short, small, nor feeble ones, but sufficiently great and strong and
armed with crooked Talons, as being the Members of Birds of Prey.

But against the Uses of several Bodies I have instanced in that refer to Man it may be objected, that these uses were not designed by Nature in the formation of the things; but that the things were by the wit of Man accommodated to those Uses.

To which I answer with Dr. *More* in the *Appendix* to his *Antidote against Atheism*. That the several useful dependencies of this kind (*viz.* of *Stones, Timber,* and *Metals* for building of Houses or Ships, the *Magnet* for Navigation, &c. Fire for melting of Metals and forging of Instruments for the purposes mentioned), we only find, not make them. For whether we think of it or no, it is, for Example, manifest, that Fuel is good to continue Fire, and Fire to melt Metals, and Metals to make Instruments to build Ships and Houses, and so on. Wherefore it being true, that there is such a subordinate usefulness in the Things themselves that are made to our Hand, it is but reason in us to impute it to such a Cause as was aware of the usefulness and serviceableness of its own Works. To which I shall add, that since we find Materials so fit to serve all the Necessities and Conveniences, and to exercise and employ the Wit and Industry of an intelligent and active. Being, and since there is such an one created that is endued with Skill and Ability to use them, and which by their help is enabled to rule over and subdue all inferiour Creatures, but without them had been left necessitous, helpless and obnoxious to Injuries above any other; and since the omniscient Creator could not but know all the Uses, to which they might and would be employed by Man, to them that acknowledge the Being of a Deity, it is little less than a Demonstration, that they were created intentionally, I do not say only, for those uses.

Methinks by all this Provision for the Use and Service of Man, the Almighty interpretatively speaks to him in this manner. I have placed thee in a spacious and well furnished World; I have endued thee with an ability of understanding what is beautiful and proportionable, and have made that which is so agreeable and delightful to thee; I have provided thee with Materials whereon to exercise and employ thy Art and Strength; I have given thee an excellent Instrument, the Hand, accommodated to make use of them all; I have distinguished the Earth into Hills, and Valleys, and Plains, and Meadows, and Woods; all these parts capable of Culture and Improvement by thy Industry; I have committed to thee for thy assistance in thy labors of Plowing, and Carrying, and Drawing, and Travel; the laborious Ox, the patient Ass, and the strong and serviceable Horse; I have created a multitude of Seeds for thee to make choice out of them, of what is most pleasant to thy Tast, and of most wholsom and pleasant Nourishment; I have also made great variety of Trees, bearing Fruit both for Food and Physick, those too capable of being meliorated and improved by Transplantation, Stercoration, Insition, Pruning, Watering, and other Arts and Devices. Till and manure thy Fields, sow them with thy Seeds, extirpate noxious and unprofitable Herbs, guard them from the invasions and spoil of Beasts, clear and fence in thy Meadows and Pastures; dress and prune thy Vines, and so rank and dispose them as is most sutable to the Climate; Plant thee Orchards, with all sorts of Fruit Trees in such order as may be most beautiful to the Eye, and most comprehensive of Plants; Gardens for culinary Herbs, and all kinds of Salletting; for delectable Flowers, to gratifie the Eye with their agreeable Colors and Figures, and thy scent with their fragrant Odors; for odoriserous and ever-green Shrubs and *Suffrutiees;* for exotick and medicinal Plants of all sorts, and dispose them in that comly order, as may be both pleasant to behold, and commodious for Access. I have furnished thee with all Materials for building, as Stone, and Timber, and Slate, and Lime, and Clay, and Earth whereof to make Bricks and Tiles. Deck and bespangle the Country with Houses and Villages convenient for thy Habitation, provided with Outhouses and Stables for the harbouring and shelter of thy Cattle, with Barns and Granaries for the reception, and custody, and sloring up thy Corn and Fruits. I have made thee a sociable Creature, for the improvement of thy Understanding by Conference, and communication of Observations and Experiments; for mutual help, assistance and defence; build thee large Towns and Cities with streight and well paved Streets, and elegant rows of Houses, adorned with magnificent Temples for my Honour and Worship, with beautiful Palaces for thy Princes and Grandees, with stately Halls for publick meetings of the Citizens and their several Companies, and the Sessions of the Courts of Judicature, besides publick Portico's and Aquæducts. I have implanted in thy Nature a desire of seeing strange and foreign and finding out unknown Countries, for the improvement and advancement of thy Knowledge in *Geography,* by observing the Bays, and Creeks, and Havens, and Promontories, the outlets of Rivers, the situations of the maritime Towns and Cities, the Longitude and Latitude, &c. of those Places: In *Politicks,* by noting their Government, their Manners, Laws and Customs, their Diet and Medicine, their Trades and Manufactures, their Houses and Buildings, their Exercises and Sports &c. In *Physiology* or Natural History, by searching out their natural Rarities, the productions both of Land and Water, what *Species* of Animals, Plants and Minerals, of Fruits and Drogues are to be found there, what Commodities for Battering and Permutation, whereby thou maist be enabled to make large Additions to Natural History, to advance those other Sciences, and to benefit and enrich thy Country by encrease of its Trade and Merchandise: I have given thee Timber and Iron to build thee Huls of Ships, tall Trees for Masts, Flax and Hemp for Sails, Cables, and Cordage for Rigging. I have armed thee with Courage and Hardiness to attempt the Seas, and traverse the spacious Plains of that liquid Element; I have assisted thee with a Compass, to direct thy Course when thou shalt be out of all Ken of Land, and have nothing in view but Sky and Water. Go thither for the Purposes before mentioned, and bring home what may be useful and beneficial to thy Country in general, or thy Self in particular.

I perswade my self, that the bountiful and gracious Author of Mans Being and Faculties, and all things else, delights in the Beauty of his Creation, and is well pleased with the Industry of Man in adorning the Earth with beautiful Cities and Castles, with pleasant Villages and Country Houses, with regular Gardens and Orchards and Plantations of all sorts of Shrubs, and Herbs, and Fruits, for Meat, Medicine or moderate Delight, with shady Woods and Groves, and Walks set with rows of elegant Trees; *with Pastures clothed with Flocks, and Valleys covered over with Corn*, and Meadows burthened with Grafs, and whatever else differeneeth a civil and well cultivated Region from a barren and desolate Wilderness.

If a Country thus planted and adorned, thus polished and civilized, thus improved to the height by all manner of Culture for the Support and Sustenance, and convenient Entertainment of innumerable multitudes of People, be not to be preferred before a Barbarous and Inhospitable *Scythia*, without Houses, without Plantations, without Corn-fields or Vineyards, where the roving *Hords* of the savage and truculent Inhabitants, transfer themselves from place to place in Wagons, as they can find Pasture and Forage for their Cattle, and live upon Milk and Flesh roasted in the Sun at the Pomels of their Saddles; or a rude and unpolished *America*, peopled with flothful and naked *Indians*, instead of well-built Houses, living in pitiful Hutts and Cabans, made of Poles set endways; then surely the brute Beasts Condition and manner of Living, to which, what we have mention'd doth nearly approach, is to be esteemed better than Mans, and Wit and Reason was in vain bestowed on him.

Lastly, I might draw an Argument of the admirable Art and Skill of the Creator and Composer of them from the incredible Smalness of some of those natural and enlivened Machines, the Bodies of Animals.

Source: Ray, John. 1979. *The Wisdom of God Manifested in the Works of the Creation, 1691.* New York & London: Garland Publishing, Inc.

Carolus Linnaeus established a classification system in order to understand the order of nature. He hoped this order would reveal the system of God's creation. At the same time, he went beyond the mere descriptions of previous naturalists in considering the role of various species within that order, as this excerpt illustrates. Linnaeus accommodated within his system the seemingly endless supply of new species discovered by exploring naturalists in his own and preceding centuries. His descriptions provided unique perspectives on the relations of animals and plants.

On the Police of Nature. (c. 1758)

CAROLUS LINNAEUS

Those laws of nature by which the number of species in the natural kingdoms is preserved undestroyed, and their relative proportions kept in proper bounds, are objects extremely worthy of our attentive pursuit and researches: in this investigation we must proceed from her simple to her more complicated institutions: the circumstance of having no predecessor in this enquiry, will, I hope, be received as a sufficient apology for venturing upon such an attempt, and for my deficiences in the execution of it. Entirely omitting the fossils, I shall endeavour to develop some of the constitutions established for this end in the animal and vegetable kingdoms.

§ 2. Vegetable Kingdom

If the many thousand species of vegetables grew together in one and the same place, some would infallibly predominate over and extirpate others; here we discover the most manifest tokens of the divine wisdom: they are separated into different parts of the world, some being natives of India, some of the temperate Zone, and others of the Polar circle. In every country different species have different stations assigned them, growing in the sea, lakes, marshes, vallies, fields, hills, rocks and shaded places, and every one has its different soil, sand, clay, earth, or chalk, allotted to it: as for example, Sweden produces about 1300 different plants; these being each confined to their proper stations, there are seldom above 50 or 100 to be found in a given place; by which it happens that no one can totally exclude another. Every plant flourishes best in its own station; if it intrude into another it is choaked up by the native species, and becoming unhealthy is at last totally devoured by the Aphides (Lice in the leaves of plants) and other insects. And that these 100 species may be the less liable to oppress one another, some of them have their time of flowering in the Spring, some in the Autumn and others in Summer.

The Animal Kingdom

§ 6. extract.

The general received opinion has been that vegetables were created for the food and uses of animals; but attending to the order of nature, we discover that animals were created upon account of plants.

§ 7

Animals serve in the first place to preserve a due proportion among vegetables: 2dly, to adorn the theatre of nature and consume every thing superfluous and useless: 3dly, to remove all impurities arising from animal and vegetable putridity; and lastly, to multiply and disseminate plants and serve them in many other respects.

§ 8

Worms first offer themselves to our consideration, but as far the greater number of these are natives of the deepest parts of the ocean, which opposes an invincible obstacle to our investigating their natural history, I am obliged to pass this division of my subject over untouched.

§ 9 and 10

The Insects are the most numerous tribe of the ministers of nature, the multitude of their species seems to vye with that of plants. . . . Some of these are always found adhering to vegetables, and subsist upon them totally when in the state of Caterpillars; such are the Papiliones (Butterfly), the Chrysomelæ (Gold-chaffer), Aphides (Leaf-louse), Cicadæ (Grasshopper), and a long catalogue besides. It is almost impossible to find a plant not exposed to the ravages of some of these, yet they are all confined to their distinct stations in the same manner as vegetables; this is proved by innumerable examples in the Pandora insectorum (Figwort wevil, Nut-wevil).

§ 35

Man is the last and highest minister of Nature, to whose use and convenience every thing is subservient; and he in many instances serves to keep up her instituted proportions: the greatest whale in the ocean, the most ferocious lion or tyger on the land, falls a victim to his boldness and address. He accomodates to his use herbs, trees, fish, birds, and every thing superfluous in the system. — Thus an exact equilibrium is kept up, where nothing is redundant or useless. But man himself is subject to this general institution, in places a-bounding with inhabitants, contagious disorders prevail with more frequency and force, and I do not know whether it be not a positive constitution of nature, that wars should be most common where men are found most numerous.

CHAPTER II

From the examples above laid down, we discover the following laws in the Police of Nature.

Vegetables, which are so many proofs of the wisdom of the great Creator, are destined always to preserve their number of species. To obtain this end, generation, nutrition, and proportion are necessary.

Animals seem created to assist in effecting these three purposes.

They assist in the generation of plants by cropping them down, and preventing a sterility, which might be the consequence of too much luxuriance: they likewise admirably contribute to their dissemination, and many of them dig up the earth for the reception of their seeds. Birds and cattle carry many of them to distant places where they never sprung before, to omit those which, after having satisfied their hunger, hoard what remains of their share in the earth: their dung likewise serves to the support of vegetables.

Their proportion seems to be among the most wonderful institutions of nature; animals seem to be the principal Ministers set over them by her to take care that of the many species of vegetables, no one should be totally extirpated: the subordination of animals, observed before, greatly assists to this end.

From the parts of Natural History already known, we may form a judgement of the importance of every one of the constitutions of nature; if the species destined to prey upon any particular animal were to perish, the greatest calamities might result from it: Nature has appointed the Quiscula to watch over the Dermestes pisorum (Pea dor), these being extirpated in North America by shooting, the peas have been totally ruined. If all the Aparrows were to be destroyed here, our plantations would be ruined by the Grylli (Cricket); America, deprived of Swine, would be infested with serpents to an intolerable degree; and we must believe the same with respect to the other servants of the Great Family of Nature, since its Author has permitted nothing to be without sufficient reason.

Very manifest indications of the wisdom of nature are to be gathered from the size of some animals being restrained in given bounds; if the Cicindela (Meadow wig), the Libellula (Dragon fly), for the Scolopendra, had been created equal in magnitude to the Lion, the whole human and animal race must have been destroyed.

One circumstance which serves to preserve the general proportion deserves our highest admiration, that the least animals are propagated in the greatest numbers, and are with the most difficulty totally extirpated: it admits no doubt that a single species of insects, though of the smallest size, can commit more ravages in a village than an Elephant; the whole species of such insects cannot be extirpated, the Elephant may be killed by a single shot.

The long lived animals propagate very slow; the class of Hawks lay but four eggs in a year, the Domestic fowl from thirty to fifty. Doves two every month; the Hare has young often in a year, and a Bee in the same period will lay 40,000 eggs.

The higher ministers of nature, the animals of prey, are slower and more indolent than others, and never seek their prey but when urged by hunger; as we observe in Lions, Tygers, and Eagles. —Beasts of prey, when they can get no dead carcasses, find considerable difficulty in getting food; and they catch generally only such of the other animals who are infirm with age or distemper; those which are young and healthy escaping them by running, flying, swimming, or in some other mode.

There are some animals likewise in the Police of Nature who are appointed as watchmen to warn other animals of their danger, as the Charadrius spinosus, Lanius (Butcher bird), Grus (Crane), Meleagris, and others, which give notice to the birds of a hawk being out in search of prey: nor would the larger conchs ever escape the Sepia loligo (Sea sleeve) unless they had the Cancer pinnotheres as a guard. Thus we see Nature resemble a well regulated state in which every individual has his proper employment and subsistence, and a proper gradation of offices and officers is appointed to correct and restrain every detrimental excess.

Source: Wilcke, Christ. Daniel. "Dissertation III. On the Police of Nature." *Amœn. Acad.* vol. 6, p. 17.

William Paley, a mathematician, theologian, and priest in the Church of England, developed the argument in this essay in order to refute the claims of atheists. He suggested that "the argument from design" clearly demonstrated the existence of an all-knowing and all-powerful creator. Recognizing that living beings are so much more complex than even the most sophisticated human creations, Paley insisted that an even greater intelligence must have designed the living world. His argument was taught in all of the universities of England in the early nineteenth century. Charles Darwin, for one, was deeply persuaded by the argument.

Natural Theology

WILLIAM PALEY (1802)

Chapter One: State of the Argument

In crossing a heath, suppose I pitched my foot against a *stone* and were asked how the stone came to be there, I might possibly answer that for anything I knew to the contrary it had lain there forever; nor would it, perhaps, be very easy to show the absurdity of this answer. But suppose I had found a *watch* upon the ground, and it should be inquired how the watch happened to be in that place, I should hardly think of the answer which I had before given, that for anything I knew the watch might have always been there. Yet why should not this answer serve for the watch as well as for the stone? Why is it not as admissible in the second case as in the first? For this reason, and for no other, namely, that when we come to inspect the watch, we perceive—what we could not discover in the stone—that its several parts are framed and put together for a purpose, e.g., that they are so formed and adjusted as to produce motion, and that motion so regulated as to point out the hour of the day; that if the different parts had been differently shaped from what they are, of a different size from what they are, or placed after any other manner or in any other order than that in which they are placed, either no motion at all would have been carried on in the machine, or none which would have answered the use that is now served by it. To reckon up a few of the plainest of these parts and of their offices, all tending to one result; we see a cylindrical box containing a coiled elastic spring, which, by its endeavor to relax itself, turns round the box. We next observe a flexible chain—artificially wrought for the sake of flexure—communicating the action of the spring from the box to the fusee. We then find a series of wheels, the teeth of which catch in and apply to each other, conducting the motion from the fusee to the balance and from the balance to the pointer, and at the same time, by the size and shape of those wheels, so regulating that motion as to terminate in causing an index, by an equable and measured progression, to pass over a given space in a given time. We take notice that the wheels are made of brass, in order to keep them from rust; the springs of steel, no other metal being so elastic; that over the face of the watch there is placed a glass, a material employed in no other part of the work, but in the room of which, if there had been any other than a transparent substance, the hour could not be seen without opening the case. This mechanism being observed—it requires indeed an examination of the instrument, and perhaps some previous knowledge of the subject, to perceive and understand it; but being once, as we have said, observed and understood—the inference we think is inevitable, that the watch must have had a maker—that there must have existed, at some time and at some place or other, an artificer or artificers who formed it for the purpose which we find it actually to answer, who comprehended its construction and designed its use.

I. Nor would it, I apprehend, weaken the conclusion, that we had never seen a watch made—that we had never known an artist capable of making one—that we were altogether incapable of executing such a piece of workmanship ourselves, or of understanding in what manner it was performed. . . .

II. Neither, secondly, would it invalidate our conclusion, that the watch sometimes went wrong or that it seldom went exactly right. The purpose of the machinery, the design, and the designer might be evident, and in the case supposed, would be evident, in whatever way we accounted for the irregularity of the movement, or whether we could account for it or not. It is not necessary that a machine be perfect in order to show with what design it was made: still less necessary, where the only question is whether it were made with any design at all.

III. Nor, thirdly, would it bring any uncertainty into the argument, if there were a few parts of the watch, concerning which we could not discover or had not yet discovered in what manner they conduced to the general effect; or even some parts, concerning which we could not ascertain whether they conduced to that effect in any manner whatever. . . .

IV. Nor, fourthly, would any man in his senses think the existence of the watch with its various machinery accounted for, by being told that it was one out of possible combinations of material forms; that whatever he had found in the place where he found the watch, must have contained some internal configuration or other; and that this configuration might be the structure now exhibited, namely, of the works of a watch, as well as a different structure.

V. Nor, fifthly, would it yield his inquiry more satisfaction, to be answered that there existed in things a principle of order, which had disposed the parts of the watch into their present form and situation. He never knew a watch made by the principle of order; nor can he even form to himself an idea of what is meant by a principle of order distinct from the intelligence of the watchmaker.

VI. Sixthly, he would be surprised to hear that the mechanism of the watch was no proof of contrivance, only a motive to induce the mind to think so.

VII. And not less surprised to be informed that the watch in his hand was nothing more than the result of the laws of *metallic* nature. It is a perversion of language to assign any law as the efficient, operative cause of any thing. A law presupposes an agent, for it is only the mode according to which an agent proceeds: it implies a power, for it is the order according to which that power acts. Without this agent, without this power, which are both distinct from itself, the *law* does nothing, is nothing. The expression, "the law of metallic nature," may sound strange and harsh to a philosophic ear; but it seems quite as justifiable as some others which are more familiar to him, such as "the law of vegetable nature," "the law of animal nature," or, indeed, as "the law of nature" in general, when assigned as the cause of phenomena, in exclusion of agency and power, or when it is substituted into the place of these.

VIII. Neither, lastly, would our observer be driven out of his conclusion or from his confidence in its truth by being told that he knew nothing at all about the matter. He knows enough for his argument; he knows the utility of the end;

he knows the subserviency and adaptation of the means to the end. These points being known, his ignorance of other points, his doubts concerning other points affect not the certainty of his reasoning. The consciousness of knowing little need not beget a distrust of that which he does know.

Chapter Two: State of the Argument Continued

Suppose, in the next place, that the person who found the watch should after some time discover that, in addition to all the properties which he had hitherto observed in it, it possessed the unexpected property of producing in the course of its movement another watch like itself—the thing is conceivable; that it contained within it a mechanism, a system of parts—a mold, for instance, or a complex adjustment of lathes, baffles, and other tools—evidently and separately calculated for this purpose. . . .

The question is not simply, How came the first watch into existence? which question, it may be pretended, is done away by supposing the series of watches thus produced from one another to have been infinite, and consequently to have had no such *first* for which it was necessary to provide a cause. This, perhaps, would have been nearly the state of the question, if nothing had been before us but an unorganized, unmechanized substance, without mark or indication of contrivance. It might be difficult to show that such substance could not have existed from eternity, either in succession—if it were possible, which I think it is not, for unorganized bodies to spring from one another—or by individual perpetuity. But that is not the question now. To suppose it to be so is to suppose that it made no difference whether he had found a watch or a stone. . . . It is in vain, therefore, to assign a series of such causes or to allege that a series may be carried back to infinity; for I do not admit that we have yet any cause at all for the phenomena, still less any series of causes either finite or infinite. Here is contrivance but no contriver; proofs of design, but no designer.

Source: Paley, William. 1800. *Natural Theology.* Edited, with an introduction by Frederick Ferré. Pp. 3–7. New York: Bobbs-Merrill.

Thomas Jefferson, author of the Declaration of Independence and third president of the United States, was also a respected observer of the natural history of the New World. He took issue, as illustrated in this excerpt, with the opinions of European naturalists who assumed that American species were smaller and less vigorous than their counterparts in other areas of the world. Jefferson relied in part on logic to dispute the authorities of Europe, and even more persuasively argued by looking directly at evidence from his surroundings in Virginia and beyond.

Notes on the State of Virginia

THOMAS JEFFERSON (1825)

Animals

Our quadrupeds have been mostly described by Linnæus and Mons. de Buffon. Of these the Mammoth, or big buffalo, as called by the Indians, must certainly have been the largest. Their tradition is, that he was carnivorous, and still exists in the northern parts of America. . . . It is well known that on the Ohio, and in many parts of America further north, tusks, grinders, and skeletons of unparalleled magnitude, are found in great numbers, some lying on the surface of the earth, and some a little below it. A Mr. Stanley, taken prisoner by the Indians near the mouth of the Tanissee, relates, that, after being transferred through several tribes, from one to another, he was at length carried over the mountains west of the Missouri to a river which runs westwardly; that these bones abounded there; and that the natives described to him the animal to which they belonged as still existing in the northern parts of their country; from which description he judged it to be an elephant. Bones of the same kind have been lately found, some feet below the surface of the earth, in salines opened on the North Holston, a branch of the Tanissee, about the latitude of $36\frac{1}{2}°$. North. From the accounts published in Europe, I suppose it to be decided, that these are of the same kind with those found in Siberia. Instances are mentioned of like animal remains found in the more southern climates of both hemispheres; but they are either so loosely mentioned as to leave a doubt of the fact, so inaccurately described as not to authorize the classing them with the great northern bones, or so rare as to found a suspicion that they have been carried thither as curiosities from more northern regions. So that on the whole there seem to be no certain vestiges of the existence of this animal further south than the salines last mentioned. It is remarkable that the tusks and skeletons have been ascribed by the naturalists of Europe to the elephant, while the grinders have been given to the hippopotamus, or riverhorse. Yet it is acknowledged, that the tusks and skeletons are much larger than those of the elephant, and the grinders many times greater than those of the hippopotamus, and essentially different in form. . . . But, 1. The skeleton of the mammoth (for so the incognitum has been called) bespeaks an animal of six times the cubic volume of the elephant, as Mons. de Buffon has admitted. The grinders are five times as large, are square, and the grinding surface studded with four or five rows of blunt points: whereas those of the elephant are broad and thin, and their grinding surface flat. 3. I have never heard an instance, and suppose there has been none, of the grinder of an elephant being found in America. 4. From the known temperature and constitution of the elephant he could never have existed in those regions where the remains of the mammoth have been found. The elephant is a native only of the torrid zone and its vicinities: if, with the assistance of warm apartments and warm clothing, he has been preserved in life in the temperate climates of Europe, it has only been for a small portion of what would have been his natural period, and no instance of his multiplication in them has ever been known. But no bones of the mammoth, as I have before observed, have been ever found further south than the salines of the Holston, and they have been found as far north as the Arctic circle. Those, therefore, who are of opinion that the elephant and mammoth are the same, must believe, 1. That the elephant known to us can exist and multiply in the frozen zone; or, 2. That an internal fire may once have warmed those regions, and since abandoned them, of which, however, the globe exhibits no unequivocal indications: or, 3. That the obliquity of the ecliptic, when these elephants lived, was so great as to include within the tropics all those regions in which the bones are found; the tropics being, as is before observed, the natural limits of habitation for the elephant. But if it be admitted that this obliquity has really decreased, and we adopt the highest rate of decrease yet pretended, that is, of one minute in a century, to transfer the northern tropic to the Arctic circle, would carry the existence of these supposed elephants 250,000 years back; a period far beyond our conception of the duration of animal bones left exposed to the open air, as these are in many instances. Besides, though these regions would then be supposed within the tropics, yet their winters would have been too severe for the sensibility of the elephant. . . . Thus nature seems to have drawn a belt of separation between these two tremendous animals, whose breadth indeed is not precisely known, though at present we may suppose it about 6½ degrees of latitude; to have assigned to the elephant the regions South of these confines, and those North to the mammoth, founding the constitution of the one in her extreme of heat, and that of the other in the extreme of cold. When the Creator has therefore separated their nature as far as the extent of the scale of animal life allowed to this planet would permit, it seems perverse to declare it the same, from a partial resemblance of their tusks and bones. But to whatever animal we ascribe these remains, it is certain such a one has existed in America, and that it has been the largest of all terrestrial beings. It should have sufficed to have rescued the earth it inhabited, and the atmosphere it breathed, from the imputation of impotence in the conception and nourishment of animal life on a large scale: to have stifled, in its birth, the opinion of a writer, the most learned too of all others in the science of animal history, that in the new world, "La nature vivante est beaucoup moins agissante, beaucoup moins forte:" that nature is less active, less energetic on one side of the globe than she is on the other. As if both sides were not warmed by the same genial sun; as if a soil of the same chemical composition, was less capable of elaboration into animal nutriment; as if the fruits and grains from that soil and sun, yielded a less rich chyle, gave less extension to the solids and fluids of the body, or produced sooner in the cartilages, membranes, and fibres, that rigidity which restrains all further extension, and terminates animal growth. The truth is, that a Pigmy and a Patagonian, a Mouse and a Mammoth, derive their dimensions from the same nutritive juices. The difference of increment depends on circumstances unsearchable to beings with our capacities. Every race of animals seems to have received from their Maker certain laws of extension at the time of their formation. Their elaborative organs were formed to produce this, while proper obstacles were opposed to its further progress. Below these limits they cannot fall, nor rise above them. What intermediate station they shall take may depend on soil, on climate, on food, on a careful choice of breeders. But all the manna of heaven would never raise the Mouse to the bulk of the Mammoth.

The opinion advanced by the Count de Buffon, is 1. That the animals common both to the old and new world, are smaller in the latter. 2. That those peculiar to the new, are on a smaller scale. 3. That those which have been domesticated in both, have degenerated in America: and 4. That on the whole it exhibits fewer species. And the reason he thinks is, that the heats of America are less; that more waters are spread over its surface by nature, and fewer of these drained off by the hand of man. In other words, that *heat* is friendly, and *moisture* adverse to the production and development of large quadrupeds. I will not meet this hypothesis on its first doubtful ground, whether the climate of America be comparatively more humid? Because we are not furnished with observations sufficient to decide this question. And though, till it be decided, we are as free to deny, as others are to affirm the fact, yet for a moment let it be supposed. The hypothesis, after this supposition, proceeds to another; that *moisture* is unfriendly to animal growth. The truth of this is inscrutable to us by reasonings a priori. Nature has hidden from us her modus agendi. Our only appeal on such questions is to experience; and I think that experience is against the supposition. It is by the assistance of *heat* and *moisture* that vegetables are elaborated from the elements of earth, air, water, and fire. We accordingly see the more humid climates produce the greater quantity of vegetables. Vegetables are mediately or immediately the food of every animal: and in proportion to the quantity of food, we see animals not only multiplied in their numbers, but improved in their bulk, as far as the laws of their nature will admit.

Source: Jefferson, Thomas. 1955. *Notes on the State of Virginia by Thomas Jefferson.* Edited with an Introduction and Notes by William Peden. Published for the Institute of Early American History and Culture at Williamsburg, Virginia by the University of North Carolina Press, Chapel Hill.

Charles Lyell produced a pioneering work when he published the three-volume Principles of Geology *excerpted here. Like the naturalists who spelled out the fundamental nature of physics and mathematics in earlier centuries, he believed that geology had underlying rules that could be observed and quantified. Lyell called his approach "uniformitarianism," suggesting that change was constant, slow, and took place over long periods of time. While such changes did not produce dramatic results on the scale of a human lifetime, all observed geological forms could be accounted for by processes still in operation, given eons of time.*

Principles of Geology

CHARLES LYELL (1830)

Geology is the science which investigates the successive changes that have taken place in the organic and inorganic kingdoms of nature; it enquires into the causes of these changes, and the influence which they have exerted in modifying the surface and external structure of our planet.

By these researches into the state of the earth and its inhabitants at former periods, we acquire a more perfect knowledge of its *present* condition, and more comprehensive views concerning the laws *now* governing its animate and inanimate productions. When we study history, we obtain a more profound insight into human nature, by instituting a comparison between the present and former states of society. We trace the long series of events which have gradually led to the actual posture of affairs; and by connecting effects with their causes, we are enabled to classify and retain in the memory a multitude of complicated relations—the various peculiarities of national character—the different degrees of moral and intellectual refinement, and numerous other circumstances, which, without historical associations, would be uninteresting or imperfectly understood. As the present condition of nations is the result of many antecedent changes, some extremely remote and others recent, some gradual, others sudden and violent, so the state of the natural world is the result of a long succession of events, and if we would enlarge our experience of the present economy of nature, we must investigate the effects of her operations in former epochs.

We often discover with surprise, on looking back into the chronicles of nations, how the fortune of some battle has influenced the fate of millions of our contemporaries, when it has long been forgotten by the mass of the population. With this remote event we may find inseparably connected the geographical boundaries of a great state, the language now spoken by the inhabitants, their peculiar manners, laws, and religious opinions. But far more astonishing and unexpected are the connexions brought to light, when we carry back our researches into the history of nature. The form of a coast, the configuration of the interior of a country, the existence and extent of lakes, valleys, and mountains, can often be traced to the former prevalence of earthquakes and volcanoes, in regions which have long been undisturbed. To these remote convulsions the present fertility of some districts, the sterile character of others, the elevation of land above the sea, the climate, and various peculiarities, may be distinctly referred.

As it is necessary that the historian and the cultivators of moral or political science should reciprocally aid each other, so the geologist and those who study natural history or physics stand in equal need of mutual assistance. A comparative anatomist may derive some accession of knowledge from the bare inspection of the remains of an extinct quadruped, but the relic throws much greater light upon his own science, when he is informed to what relative era it belonged, what plants and animals were its contemporaries, in what degree of latitude it once existed, and other historical details. A fossil shell may interest a conchologist, though he be ignorant of the locality from which it came; but it will be of more value when he learns with what other species it was associated, whether they were marine or fresh-water, whether the strata containing them were at a certain elevation above the sea, and what relative position they held in regard to other groups of strata, with many other particulars determinable by an experienced geologist alone. On the other hand, the skill of the comparative anatomist and conchologist are often indispensable to those engaged in geological research, although it will rarely happen that the geologist will himself combine these different qualifications in his own person.

It was long ere the distinct nature and legitimate objects of geology were fully recognized, and it was at first confounded with many other branches of inquiry, just as the limits of history, poetry, and mythology were ill-defined in the infancy of civilization. Werner appears to have regarded geology as little other than a subordinate department of mineralogy, and Desmarest included it under the head of Physical Geography. But the identification of its objects with those of Cosmogony has been the most common and serious source of confusion. The first who endeavoured to draw a clear line of demarcation between these distinct departments was Hutton, who declared that geology was in no ways concerned "with questions as to the origin of things." But his doctrine on this head was vehemently opposed at first, and although it has gradually gained ground, and will ultimately prevail, it is yet far from being established. We shall attempt in the sequel of this work to demonstrate that geology differs as widely from cosmogony, as speculations concerning the creation of man differ from history. But before we enter more at large on this controverted question, we shall endeavour to trace the progress of opinion on this topic, from the earliest ages, to the commencement of the present century.

We have considered, in the preceding chapters, many of the most popular grounds of opposition to the doctrine, that all former changes of the organic and inorganic creation are referrible to one uninterrupted succession of physical events, governed by the laws now in operation.

As the principles of the science must always remain unsettled so long as no fixed opinions are entertained on this fundamental question, we shall proceed to examine other objections which have been urged against the assumption of uniformity in the order of nature.

Now, in the first place, we may observe, that many naturalists have been guilty of no small inconsistency in endeavouring to connect the phenomena of the earliest vegetation with a nascent condition of organic life, and at the same time to deduce, from the numerical predominance of certain types of form, the greater heat of the ancient climate. The arguments in favour of the latter conclusion are without any force, unless we can assume that the rules followed by the Author of Nature in the creation and distribution of organic beings were the same formerly as now; and that as certain families of animals and plants are now most abundant, or exclusively confined to regions where there is a certain temperature, a certain degree of humidity, intensity of light, and other conditions, so also the same phenomena were exhibited at every former era. If this postulate be denied, and the prevalence of particular families be declared to depend on a certain order of precedence in the introduction of different classes into the earth, and if it be maintained that the standard of organization was raised successively, we must then ascribe the numerical preponderance in the earlier ages of plants of simpler structure, *not to the heat,* but to those different laws which regulate organic life in newly created worlds. If, according to the laws of progressive development, cryptogamic plants always flourish for ages before the dicotyledonous order can be established, then is the small proportion of the latter fully explained; for in this case, whatever may have been the mildness or severity of the climate, they could not make their appearance. Before we can infer an elevated temperature in high latitudes, from the presence of arborescent Ferns, Lycopodiaceæ, and other allied families, we must be permitted to assume, that at all times, past and future, a heated and moist atmosphere pervading the northern hemisphere has a tendency to produce in the vegetation a predominance of analogous types of form. We grant, indeed, that there may be a connexion between an extraordinary profusion of monocotyledonous plants, and a youthful condition of the world, if the dogma of certain cosmogonists be true, that planets, like certain projectiles, are always red hot when they are first cast; but to this arbitrary hypothesis we need not again revert.

Source: Lyell, Charles. 1830. *Principles of Geology, With a new introduction by Martin J. S. Rudwick.* Volume I. Chicago and London: The University of Chicago Press.

The state of Massachusetts commissioned George Emerson to write a report on the state's trees and shrubs. In his account of the forests and woody resources of the state, Emerson chronicled an array of facts with economic and practical implications. In addition, he included more trivial details that highlighted the unique character of human interest in natural history in that period. Of special interest in this excerpt is the way Emerson questioned the effects of deforestation on the long-term health of the landscape.

A Report on the Trees and Shrubs Growing Naturally in the Forests of Massachusetts

GEORGE EMERSON (1850)

1. Forests create or gradually but constantly improve a soil. The roots penetrate deeply into the ground, and thus let in the air to produce its slow but sure effects. The radicles decompose the grains of sand, and extract from them some of the elements essential to a soil; they drink in moisture and the carbonic acid which has been formed beneath, or brought down from the atmosphere above, the surface; and from these several elements, acted on by heat, light and air, in the leaves, and by that unknown influence, vegetable life, are formed the various substances which compose the plant. The annual deposit of leaves, and the final decay of the branches and trunk, go to constitute the mould upon which other plants grow. And the soil thus formed is kept by the thick matting of the roots from washing away.

An unprotected hill soon loses its soil. Every rain bears away a portion, till it becomes a bare rock, and the slow process must recommence by which rock had been originally converted into soil. That process takes place slowly on all uncovered rocks, but far more surely and rapidly under cover of trees. There also the invisible sporule, borne thither on the wind, perhaps, from a distant continent, attaches itself to the naked rock and vegetates; encrusting its surface with a lichen which gets thence a foothold and an alkali, while it lives on the atmosphere. From the thin layer left by its decay, another species springs, which in turn gives place to mosses and herbaceous plants. Whoever has climbed Monument Mountain in Stockbridge, has had an opportunity of observing this process in its different stages and circumstances. On the projecting cliffs of white quartz, of which the mountain consists, the beautiful lichens which paint its sides have made no more progress than if the mountain had been thrown up two years ago. They are spread upon it as thin as paper, and perfectly fresh. Wherever they decay, the violence of the rain and winds washes them clean off, and the work is begun each year anew. But in the protected crevices, and under shelter of the few trees and shrubs that have found root-hold there, a soil is forming or is already formed. What happens here takes place on all mountain tops in New England. A sheltering tree allows the creative action to take place.

2. Another use of forests is to serve as conductors of electricity between the clouds and its great reservoir the earth; thus giving activity to the vital powers of plants, and leading the clouds to discharge their contents upon the earth. A few tall trees on the summit of a hill are sufficient to produce this effect. A charged thunder cloud, which passes unbroken over a bare hill, will pour down its moisture, if its electricity is drawn off by these natural conductors. The dry sterility of some parts of Spain, anciently very fertile, is probably owing, in a great degree, to the improvident destruction of the forests, and the absurd laws which discourage their renewal. The forests also coat the earth and keep it warm in winter, shutting in the central heat which would otherwise more rapidly radiate into space and be lost. If you go into the woods at the end of a severe winter, you may any where easily drive down a stake without impediment from the frost; while, in the open field by their edge, you find a foot or more of earth frozen solid. Forests act not less favorably as a protection against the excessive heat of the summer's sun, which rapidly evaporates the moisture and parches up the surface. The first mahogany cutters in Honduras found the cold under the immense forests so great, that they were obliged, though within 16° of the equator, to kindle fires to keep themselves warm. The rain, falling on the woods of a hill side, is retained by the deep and spongy mass formed by the roots and the accumulated deposit of leaves, instead of rushing down, as it otherwise would, in torrents, carrying with it great quantities of loose soil. Protected also from rapid evaporation, it remains laid up as in a reservoir, trickling gradually out and forming perennial streams, watering and fertilizing the low country through the longest summers, and moderating the violence of droughts by mists and dews. All along the coast of New England, numerous little streams, which were formerly fed by the forests, and often rolled a volume of water sufficient to turn a mill in summer, are now dried up at that season, and only furnish a drain for the melting snows of spring, or the occasional great rains of autumn.

Forests thus equalize the temperature and soften the climate, protecting from the extremes of cold and heat, dryness and humidity. There is little doubt that, if the ancient forests of Spain could be restored to its hills, its ancient fertility would return. Now, there is nothing to conduct electricity, nothing to arrest the clouds and make them pour their treasures upon the earth, no reservoirs to lay up the winter's rain in store against the droughts of summer.

Such are some of the suggestions which I have desired to lay before my fellow-citizens of Massachusetts, for the improvement of their forests and the redemption of their waste lands. I have opened, very imperfectly, the great and important study of the history and management of forest trees. A tree is the most magnificent among the material works of God. The nature, the relations to soil, to climate and to exposure, the affinities, the properties and the uses to man and other animals, the dangers from enemies and diseases within and without, and the circumstances necessary to secure the health, growth and beauty of the trees of any one family, are subjects worthy of the deliberate and mature and long continued attention of any man, of whatever intelligence, and with whatever resources of science. The best disposition of trees in the landscape, the treatment of each according to its character and appearance at all seasons of the year, so as to foresee and to produce the desired effect at every point which the eye can reach, and the adaptation of the various kinds of trees to the houses, churches, bridges, and other structures already existing or to be erected, and also to water, and to roads—things evidently possible and yet indefinitely difficult—to do all this successfully is the province of an art, which well deserves to take its place in the front rank among the fine arts; whether we consider the science, taste and skill which it calls into play, the vastness of the scale on which it acts, or the grandeur of the end which it has in view.

But why should it be thought important to reclaim or render valuable the waste or worthless lands of Massachusetts? There are millions of acres of land in the Western States far richer than any in our State, which may be purchased for much less than it will cost to render barren land productive. Why not go thither and occupy the rich wild lands? For many reasons. This is our native land. It is painful to break the chain of affection which connects us with it. It is painful to separate members of the same family. Every improvement in agriculture, in the management of the forests, and in the use of the other natural resources of our State, makes it capable of sustaining a larger population, and thus enables more of our young men and young women to remain with us, rendering home dearer to those who would otherwise be left behind. The advantages of our life, in the long settled parts of the Bay State, are greater than can be expected, for more than a single generation to come, in the newly settled regions of the valley of the Mississippi or in any other new region. There are still higher reasons. We live in a climate and on a soil, best adapted, from their very severity and sterility, to bring out the energies of mind and body, and to form a race of hardy and resolute men. We have our churches, our schools, our libraries, our intelligent and virtuous neighbors—dearer to us than any strangers can be. These we are not willing to leave. We wish that our children should grow up under the influence of the institutions which our forefathers have formed and left to us, and which we have been endeavoring to improve. Here we wish to live and to die; and when we die, we wish to be surrounded by those who are most dear to us.

Source: Emerson, George. 1850. *A Report on the Trees and Shrubs Growing Naturally in the Forests of Massachusetts: Published agreeably to an order of the legislature by the commissioners of the zoological and botanical survey of the state.* Boston: Charles C. Little & James Brown.

Henry David Thoreau spent a year living on the outskirts of Concord, Massachusetts, in a shack at the edge of Walden Pond. He hoped to examine his relationship with nature and human society by moving closer to the edge of the pond. Thoreau recorded his observations in a journal that has become a classic in American natural history literature. His reflections set a tone for nature essays ever since. As illustrated here, description blends with philosophical views that often put aesthetics ahead of economics.

Walden

HENRY DAVID THOREAU (1854)

I went to the woods because I wished to live deliberately, to front only the essential facts of life, and see if I could not learn what it had to teach, and not, when I came to die, discover that I had not lived. I did not wish to live what was not life, living is so dear; nor did I wish to practise resignation, unless it was quite necessary. I wanted to live deep and suck out all the marrow of life, to live so sturdily and Spartan-like as to put to rout all that was not life, to cut a broad swath and shave close, to drive life into a corner, and reduce it to its lowest terms, and, if it proved to be mean, why then to get the whole and genuine meanness of it, and publish its meanness to the world; or if it were sublime, to know it by experience, and be able to give a true account of it in my next excursion. For most men, it appears to me, are in a strange uncertainty about it, whether it is of the devil or of God, and have somewhat hastily concluded that it is the chief end of man here to "glorify God and enjoy him forever."

Still we live meanly, like ants; though the fable tells us that we were long ago changed into men; like pygmies we fight with cranes; it is error upon error, and clout upon clout, and our best virtue has for its occasion a superfluous and evitable wretchedness. Our life is frittered away by detail. An honest man has hardly need to count more than his ten fingers, or in extreme cases he may add his ten toes, and lump the rest. Simplicity, simplicity, simplicity! I say, let your affairs be as two or three, and not a hundred or a thousand; instead of a million count half a dozen, and keep your accounts on your thumb-nail. In the midst of this chopping sea of civilized life, such are the clouds and storms and quicksands and thousand-and-one items to be allowed for, that a man has to live, if he would not founder and go to the bottom and not make his port at all, by dead reckoning, and he must be a great calculator indeed who succeeds. Simplify, simplify. Instead of three meals a day, if it be necessary eat but one; instead of a hundred dishes, five; and reduce other things in proportion. . . .

I perceive that we inhabitants of New England live this mean life that we do because our vision does not penetrate the surface of things. We think that that is which appears to be. If a man should walk through this town and see only the reality, where, think you, would the "Mill-dam" go to? If he should give us an account of the realities he beheld there, we should not recognize the place in his description. Look at a meeting-house, or a court-house, or a jail, or a shop, or a dwelling-house, and say what that thing really is before a true gaze, and they would all go to pieces in your account of them. Men esteem truth remote, in the outskirts of the system, behind the farthest star, before Adam and after the last man. In eternity there is indeed something true and sublime. But all these times and places and occasions are now and here. God himself culminates in the present moment, and will never be more divine in the lapse of all the ages. And we are enabled to apprehend at all what is sublime and noble only by the perpetual instilling and drenching of the reality that surrounds us. The universe constantly and obediently answers to our conceptions; whether we travel fast or slow, the track is laid for us. Let us spend our lives in conceiving then. The poet or the artist never yet had so fair and noble a design but some of his posterity at least could accomplish it.

Let us spend one day as deliberately as Nature, and not be thrown off the track by every nutshell and mosquito's wing that falls on the rails. Let us rise early and fast, or break fast, gently and without perturbation; let company come and let company go, let the bells ring and the children cry—determined to make a day of it. Why should we knock under and go with the stream? Let us not be upset and overwhelmed in that terrible rapid and whirlpool called a dinner, situated in the meridian shallows. Weather this danger and you are safe, for the rest of the way is down hill. With unrelaxed nerves, with morning vigor, sail by it, looking another way, tied to the mast like Ulysses. If the engine whistles, let it whistle till it is hoarse for its pains. If the bell rings, why should we run? We will consider what kind of music they are like. Let us settle ourselves, and work and wedge our feet downward through the mud and slush of opinion, and prejudice, and tradition, and delusion, and appearance, that alluvion which covers the globe, through Paris and London, through New York and Boston and Concord, through Church and State, through poetry and philosophy and religion, till we come to a hard bottom and rocks in place, which we can call reality, and say, This is, and no mistake; and then begin, having a point d'appui, below freshet and frost and fire, a place where you might found a wall or a state, or set a lamp-post safely, or perhaps a gauge, not a Nilometer, but a Realometer, that future ages might know how deep a freshet of shams and appearances had gathered from time to time. If you stand right fronting and face to face to a fact, you will see the sun glimmer on both its surfaces, as if it were a cimeter, and feel its sweet edge dividing you through the heart and marrow, and so you will happily conclude your mortal career. Be it life or death, we crave only reality. If we are really dying, let us hear the rattle in our throats and feel cold in the extremities; if we are alive, let us go about our business.

Time is but the stream I go a-fishing in. I drink at it; but while I drink I see the sandy bottom and detect how shallow it is. Its thin current slides away, but eternity remains. I would drink deeper; fish in the sky, whose bottom is pebbly with stars. I cannot count one. I know not the first letter of the alphabet. I have always been regretting that I was not as wise as the day I was born. The intellect is a cleaver; it discerns and rifts its way into the secret of things. I do not wish to be any more busy with my hands than is necessary. My head is hands and feet. I feel all my best faculties concentrated in it. My instinct tells me that my head is an organ for burrowing, as some creatures use their snout and fore paws, and with it I would mine and burrow my way through these hills. I think that the richest vein is somewhere hereabouts; so by the divining-rod and thin rising vapors I judge; and here I will begin to mine.

Source: Thoreau, Henry David. 1854. *Walden.* Pp. 85–86, 91–93.

Charles Darwin, in his book On the Origin of Species *excerpted here, spelled out a mechanism for evolution, providing a basis for understanding biological change. Although Darwin was justified in his fear that the book would ignite a controversy over human origins, he did not discuss human evolution there and avoided challenges from critics by remaining secluded at his home outside London. The book consists primarily of evidence from his extensive studies of biological forms and analogies to domestic breeding of plants and animals.*

On the Origin of Species

CHARLES DARWIN (1859)

Summary of Chapter.—If during the long course of ages and under varying conditions of life, organic beings vary at all in the several parts of their organisation, and I think this cannot be disputed; if there be, owing to the high geometrical powers of increase of each species, at some age, season, or year, a severe struggle for life, and this certainly cannot be disputed; then, considering the infinite complexity of the relations of all organic beings to each other and to their conditions of existence, causing an infinite diversity in structure, constitution, and habits, to be advantageous to them, I think it would be a most extraordinary fact if no variation ever had occurred useful to each being's own welfare, in the same way as so many variations have occurred useful to man. But if variations useful to any organic being do occur, assuredly individuals thus characterised will have the best chance of being preserved in the struggle for life; and from the strong principle of inheritance they will tend to produce offspring similarly characterised. This principle of preservation, I have called, for the sake of brevity, Natural Selection. Natural selection, on the principle of qualities being inherited at corresponding ages, can modify the egg, seed, or young, as easily as the adult. Amongst many animals, sexual selection will give its aid to ordinary selection, by assuring to the most vigorous and best adapted males the greatest number of offspring. Sexual selection will also give characters useful to the males alone, in their struggles with other males.

Whether natural selection has really thus acted in nature, in modifying and adapting the various forms of life to their several conditions and stations, must be judged of by the general tenour and balance of evidence given in the following chapters. But we already see how it entails extinction; and how largely extinction has acted in the world's history, geology plainly declares. Natural selection, also, leads to divergence of character; for more living beings can be supported on the same area the more they diverge in structure, habits, and constitution, of which we see proof by looking at the inhabitants of any small spot or at naturalised productions. Therefore during the modification of the descendants of any one species, and during the incessant struggle of all species to increase in numbers, the more diversified these descendants become, the better will be their chance of succeeding in the battle of life. Thus the small differences distinguishing varieties of the same species, will steadily tend to increase till they come to equal the greater differences between species of the same genus, or even of distinct genera.

We have seen that it is the common, the widely-diffused, and widely-ranging species, belonging to the larger genera, which vary most; and these will tend to transmit to their modified offspring that superiority which now makes them dominant in their own countries. Natural selection, as has just been remarked, leads to divergence of character and to much extinction of the less improved and intermediate forms of life. On these principles, I believe, the nature of the affinities of all organic beings may be explained. It is a truly wonderful fact—the wonder of which we are apt to overlook from familiarity—that all animals and all plants throughout all time and space should be related to each other in group subordinate to group, in the manner which we everywhere behold—namely, varieties of the same species most closely related together, species of the same genus less closely and unequally related together, forming sections and sub-genera, species of distinct genera much less closely related, and genera related in different degrees, forming sub-families, families, orders, sub-classes, and classes. The several subordinate groups in any class cannot be ranked in a single file, but seem rather to be clustered round points, and these round other points, and so on in almost endless cycles. On the view that each species has been independently created, I can see no explanation of this great fact in the classification of all organic beings; but, to the best of my judgment, it is explained through inheritance and the complex action of natural selection, entailing extinction and divergence of character, as we have seen illustrated in the diagram.

The affinities of all the beings of the same class have sometimes been represented by a great tree. I believe this simile largely speaks the truth. The green and budding twigs may represent existing species; and those produced during each former year may represent the long succession of extinct species. At each period of growth all the growing twigs have tried to branch out on all sides, and to overtop and kill the surrounding twigs and branches, in the same manner as species and groups of species have tried to overmaster other species in the great battle for life. The limbs divided into great branches, and these into lesser and lesser branches, were themselves once, when the tree was small, budding twigs; and this connexion of the former and present buds by ramifying branches may well represent the classification of all extinct and living species in groups subordinate to groups. Of the many twigs which flourished when the tree was a mere bush, only two or three, now grown into great branches, yet survive and bear all the other branches; so with the species which lived during long-past geological periods, very few now have living and modified descendants. From the first growth of the tree, many a limb and branch has decayed and dropped off; and these lost branches of various sizes may represent those whole orders, families, and genera which have now no living representatives, and which are known to us only from having been found in a fossil state. As we here and there see a thin straggling branch springing from a fork low down in a tree, and which by some chance has been favoured and is still alive on its summit, so we occasionally see an animal like the Ornithorhynchus or Lepidosiren, which in some small degree connects by its affinities two large branches of life, and which has apparently been saved from fatal competition by having inhabited a protected station. As buds give rise by growth to fresh buds, and these, if vigorous, branch out and overtop on all sides many a feebler branch, so by generation I believe it has been with the great Tree of Life, which fills with its dead and broken branches the crust of the earth, and covers the surface with its ever branching and beautiful ramifications.

Source: Darwin, Charles. 1859. *On the Origin of Species by Charles Darwin.* Pp. 126–130.

Vernon Bailey worked as a government biologist throughout his long career, from the 1880s to the 1940s. Much of his early work involved the tracking and trapping of small mammals for collections in the Smithsonian Institution. By the first decade of the twentieth century, he participated actively in the destruction of "predatory animals," a term that came into common use only in this context at the time. In this article, he reported on the various successes of government bounty hunters in removing wolves and other livestock killers from the expanding grazing ranges of the West.

Destruction of Wolves and Coyotes: Results Obtained During 1907. Bureau of Biological Survey—Circular No. 63.

VERNON BAILEY (1908)

Wolves and coyotes cause a loss to the stockmen and farmers of the United States of several millions of dollars annually, and in some of the Northern States they threaten the extermination of deer on many of the best hunting grounds.

These losses can be prevented only by intelligent and concerted action throughout the wolf-infested country, and the Biological Survey aims to furnish information that will aid in securing the best results in the war against these pests. Early in 1907 a bulletin and two circulars on wolves were widely distributed in the region where the animals are most destructive. These publications contain brief records of losses from wolves, and directions for finding the dens and capturing the pups, for trapping and poisoning the old wolves, and for building wolf-proof fences. A year has passed since the publications were distributed, and while complete returns giving total results are not at hand, the records received indicate a marked increase in the number of wolves destroyed.

Bounty Records

Wyoming is the only wolf-infested State from which satisfactory bounty records are obtainable. During the year ending October 1, 1906, or previous to the distribution of the wolf circulars, approximately 1,607 wolves were presented for bounty. During the year beginning March 1, 1907, or after the wolf circulars were distributed, 2,035 wolves were presented, showing an increase over the preceding year of 328.

Wolves and Coyotes Killed in and Near National Forests in 1907

The Forest Service has made vigorous efforts to destroy wolves and other predatory animals on and near the national forests, and through its force of forest rangers has carried on the most systematic and successful war on these pests ever undertaken. Besides the regular force of rangers a number of expert hunters and trappers have been employed in the worst infested regions and gratifying results have been obtained. It must be borne in mind, however, that the areas thus protected are but widely separated spots in a vast extent of wolf country, and unless ranchmen and settlers are stimulated to similar efforts permanent results are not to be expected.

Following is a record of wolves and coyotes killed in 1907, furnished by the Forest Service. In many cases the records are incomplete or approximate, but they come from widely scattered localities and serve to give an idea of the success of the war against these animals. Numerous bears, mountain lions, bobcats, and other animals also were killed. In making up these reports the supervisors of the various national forests have added to the number of animals killed by forest rangers and hunters employed by the Forest Service those killed in the vicinity of the forests by ranchmen, cowboys, and professional hunters, and in many cases it has been impossible to separate the records. Also in a few cases bounty records for the county have been included. Hence the Forest Service should not be credited with the total number of animals killed.

The capture of 1,723 wolves is reported from 39 national forests, which comprise an area of 72,760 square miles. The surrounding country included may amount to as much more, making an area of about 145,520 square miles to which the reports relate, or about one-tenth of the total area inhabited by wolves in the United States.

The capture of 23,208 coyotes is reported from 77 national forests, which comprise an area of 106,746 square miles. This if doubled to include the surrounding country to which the reports also refer makes an area of approximately 213,492 square miles, or about one-ninth of that inhabited by coyotes in the United States.

The following directions for destroying wolves are republished, with slight changes, from Circular 55 of the Biological Survey.

Capture of Pups

By the destruction of the young the increase of wolves and coyotes may be prevented more effectually and economically than by any other way. The large size of the litters makes the method especially important, as wolves usually have 6 to 10 pups, and coyotes 5 to 9. It is now positively known that both wolves and coyotes pair for the breeding season, and that the males stay with the females and help feed and care for the pups during the summer. Wolves and probably coyotes do not breed until 2 years old, which accounts for the presence of roving bands during the breeding season.

Trapping

For wolves the best No. 4 double-spring trap with heavy welded or special wolf chain should be used. If the trap is to be fastened to a stationary object, the chain should have a swivel at each end. If to a drag, one swivel next to the trap is enough. Always use a drag if possible. The best is a stone of 30 or 40 pounds weight, to which the chain is securely wired. A long oval stone is best. A piece of telegraph wire or smooth fence wire 5 or 6 feet long should be passed around one end of the stone; then doubled through the trap ring, with a twist to hold the ring in the middle; then around the other end of the stone and back on the opposite side to connect with the first loop. If properly fastened, a jerk on the trap tends to draw together and tighten the loops, and the spring of the connecting wire prevents a sudden jar that might break trap or chain. If an oval stone is not at hand, a triangular or square stone may be used by passing the wire over the three or four sides and securely connecting it above and below.

If no stones are to be had and it is necessary to stake the traps, twisted iron stakes that can be driven below the surface of the ground should be used. They should be of good iron straps, at least. 18 inches long, three-fourths of an inch wide, and three-sixteenths of an inch thick, turned over at the top into a P-shaped loop to connect with the ring of the trap chain.

When possible, place the trap between two tuffs of grass or weeds, so it

WOLVES AND COYOTES KILLED IN OR NEAR NATIONAL FORESTS IN 1907.

[From supervisors' reports to Forest Service.]

Forest	Wolves	Coyotes	Remarks
WYOMING.			
Yellowstone	79	218	Old wolves, 8; pups, 71.
Bear Lodge	925	1,165	From Crook County records.
Medicine Bow and Sierra Madre	5	600	
MONTANA.			
Highwood	22	79	Wolf pups, 16; coyote pups, 45.
Little Belt	—	24	
Big Hole	30	300	
Gallatin	15	595	From bounty records of Gallatia County.
Otter	60	225	Including 51 wolf pups.
Kootenai and Cabinet	11	141	
Hellgate	—	248	
Lewis and Clark	—	74	
Helena	50	550	Numbers estimated.
Absaroka and Crazy Mountain	73	393	
IDAHO.			
Sawtooth and Payette	6	1,884	
Port Neuf	—	93	
Pocatello	—	45	
Raft River	—	420	
Cassia	—	640	
Bitter Root	—	18	
Weiser	—	315	
Caribou	—	306	
Salmon River	8	160	
UTAH.			
Dixie	—	188	
Sevier	—	1,565	
La Sal and Monticello	—	160	From county records.
Fish Lake	—	644	
Beaver	—	428	
Aquarius	—	60	
Fillmore	—	1,429	
Manti	—	405	
Uinta	—	122	
COLORADO.			
Medicine Bow	13	950	Of the wolves, 7 were pups.
San Juan	1	365	
Holy Cross	—	294	
Pikes Peak	—	346	
Wet Mountain and San Isabel	—	142	
Montezuma	6	55	
White River	45	596	
Gunnison	—	254	
OKLAHOMA.			
Wichita	3	15	

WOLVES AND COYOTES KILLED IN OR NEAR NATIONAL FORESTS IN 1907.

[From supervisors' reports to Forest Service.]

Forest	Wolves	Coyotes	Remarks
NEW MEXICO.			
Gila	[a]75	242	
Magdalena	6	51	
San Mateo	6	9	
Sacramento	22	50	
Manzano	7	81	
Lincoln	76	77	
Mount Taylor	—	34	
Gallinas	40	—	
ARIZONA.			
Santa Rita and Dragoon	45	471	
Chiricahua	30	198	
Huachuca and Tumacacori	14	25	Presented for bounty in Santa Cruz County.
Tonto	7	31	
Prescott	11	120	Presented for bounty in Yavapai County.
San Francisco	16	274	Presented for bounty in Coconino County.
Grand Canyon	4	305	
CALIFORNIA.			
Lassen Peak	—	26	
Klamath	—	80	
Tahoe	—	6	
Inyo and Sierra	—	112	
San Luis Obispo	—	10	
NEVADA.			
Tolyabe, Toquima, and Monitor	—	500	Approximately.
OREGON.			
Imnaha	—	914	
Blue Mountain	—	200	
Goose Lake and Fremont	2	2,150	
Heppner	—	26	
WASHINGTON.			
Wenaha	—	300	
Washington	4	375	
Olympic	6	—	
RECAPITULATION BY STATES			
Wyoming	1,009	1,983	
Montana	261	2,629	
Idaho	14	3,881	
Washington	10	675	
Colorado	65	2,362	
Oklahoma	3	15	
New Mexico	232	544	
Arizona	127	1,424	
Utah	—	5,001	
Nevada	—	500	
California	—	224	
Oregon	2	3,290	
TOTAL	**1,723**	**23,208**	

[a] On the Gila National Forest 36 wolves and 30 coyotes were killed by one forest guard, who sent the skulls to the Biological Survey for identification, as well as the skulls of 9 bears, 7 mountain lions, 17 bobcats, and 46 gray foxes. One den of 8 very young wolf pups was taken March 13.

can be readily approached from one side only. Bury the stone, chain, and trap out of sight, with the trap nearest to the runway where the wolves follow a trail or road, cross a narrow pass, or visit a carcass. The trap should be flush with the surface of the ground and the jaws and pan covered with a piece of paper to keep the earth from clogging under the pan. Fine earth should then be sprinkled over the paper until all traces of trap and paper are concealed. The surface of the ground and surroundings should appear as nearly as possible undisturbed. The dust may be given a natural appearance by sprinkling it with water. Touching the ground or other objects with the hands, spitting near the trap, or in any way leaving a trace of human odors near by should be avoided. Old, well-scented gloves should be worn, and a little of the scent used for the traps should be rubbed on the shoe soles. A piece of old cowhide may be used to stand on and to pile the loose earth on while burying the drag and trap.

For coyotes use the best No. 3 double-spring trap, unless in a wolf country, where it is better to use a trap strong enough to hold a wolf. In setting the trap use the same method and bait as for wolves, but the traps may be staked or fastened to a stationary object with more safety.

Use of Scents

Success in trapping depends largely on the use of a scent that will attract wolves and coyotes to the traps and keep them tramping and pawing there until caught. Meat bait alone is of little use, and often, indeed, scares the animals away. Of the many scents and combinations tested the fetid bait has proved most successful.

Fetid bait—Place half a pound of raw beef or venison in a wide-mouthed bottle and let it stand in a warm place (but not in the sun) for two to six weeks, or until it is thoroughly decayed and the odor has become as offensive as possible. When decomposition has reached the proper stage, add a quart of sperm oil or any liquid animal oil. Lard oil may be used, but prairie-dog oil is better. Then add 1 ounce of pulverized asafetida and 1 ounce of tincture of Siberian musk or Tonquin musk. If this can not be procured, use in its place 1 ounce of dry, pulverized castoreum (beaver castor) or 1 ounce of the common musk sold for perfumery. Mix well and bottle securely until used.

After setting the trap, apply the scent with a stick or straw or by pouring from the bottle to the grass, weeds, or ground on the side of the trap opposite that from which the wolf would naturally approach. Never put scent on the trap, as the first impulse of the wolf after snifling the scent is to roll on it.

This bait is very attractive also to cattle and horses, which are sure to tramp over and paw out the traps if set where they can be reached.

Poisoning

No poison has yet proved so effective as pure sulphate of strychnine, provided the proper dose is used. The most effective dose is 4 grains for wolves and 2 grains for coyotes. The common 3-grain gelatin capsules sold by druggists will hold, if well filled, 4 grains of strychnine and are better than the larger capsules. The regular 2-grain capsules should be used for coyotes. The capsules should be filled, securely capped, and every trace of the intensely bitter drug wiped from the outside.

Each capsule should be inserted in a piece of beef suet the size of a walnut and the cavity securely closed, to keep out moisture. Lean meat should not be used, as the juice soon dissolves the gelatin of the capsule. The necessary number of poisoned baits may be prepared and carried in a tin can or pail. They should never be handled except with gloved hands or forceps. The baits may be dropped from horseback along a scented drag line made by dragging an old bone or piece of hide well saturated with the fetid scent, or they may be placed around or partly under any carcass on which the wolves or coyotes are feeding, or along trails which they are in the habit of following.

Gelatin capsules quickly dissolve in the juices of the stomach. Strychnine taken on an empty stomach sometimes kills in a very few minutes, but on a full stomach its action is much slower, and the animal may have time to travel a considerable distance.

Wolf and Coyote Proof Fences

Under present conditions it is entirely impracticable to fence any great part of the western stock range against wolves and coyotes, but in many cases limited areas may be fenced to advantage. In sections where cattle are fed during the winter months wolves often kill them on the open feeding grounds. These, as well as small home pastures, for both cattle and sheep, may be inclosed with wolf and coyote proof fences at relatively slight cost, often less than the value of the stock killed during one season.

Source: Bailey, Vernon. 1907. *Destruction of Wolves and Coyotes: Results Obtained During 1907.* United States Department of Agriculture, Bureau of Biological Survey—Circular No. 63.

Edward A. Goldman played a leading role in natural history studies and biological field research for the U.S. government. By the early 1920s, he was working to expand government funding for this research. A significant component of that work included the destruction of predatory animals on behalf of western ranchers. As demonstrated here, Goldman articulated an argument that the "balance of nature" had been upset by decades of human settlement, and that to stop killing predators would do nothing to restore the balance. He advocated continued extermination and insisted that continued funding was needed for that work as well as other research that would improve access to biological resources.

The Predatory Mammal Problem and the Balance of Nature

EDWARD A. GOLDMAN (1924)

The treatment that should be accorded certain of the larger predatory mammals—whether their depredations should be tolerated, or whether extermination in certain areas should become a fixed policy—is a problem concerning which there is some difference of opinion. This problem is intimately bound up with complex relationships existing in nature, some of which must receive consideration in order that the problem in its true bearings can properly be understood.

Under normal primeval conditions there is a tendency everywhere to establish a nearly balanced status through a continuous process of readjustment in the contacts existing between species. This somewhat hypothetical condition is commonly referred to as the balance of nature, the biological balance, or the climax association. By it is meant that through competition in the struggle for existence a point has been approached where none of the species in a given area is increasing or decreasing in number of individuals. In this conception animals and plants are closely associated, and both kingdoms therefore are included within its scope. Each species has certain imperfectly known environmental requirements, or meets pressure from other species which under normal conditions definitely limit its numbers and fix its geographic distribution.

Some of the factors preventing expansion beyond the borders of definite areas may be physical barriers, improper temperatures, too much or too little moisture, or (for animals) lack of suitable vegetable food or other animals upon which to prey, or the species may itself be preyed upon to a degree that renders its existence impossible beyond certain limits. Many species when fortuitously or through human agency purposely introduced into new areas not formerly accessible, as, for example rabbits and such in Australia, mongooses and English sparrows in America, may multiply amazingly owing to abundance of food, absence of natural enemies, or other favoring conditions. The effect of such introductions on native species is often disastrous because of the new competition that must be met, and a new alignment of relationships ensues. Leaving man out of the setting as a disturbing factor, natural causes, such as fire, drought, and floods, epizooties, or conditions temporarily favoring the expansion of certain species at the expense of others, frequently upset the local balance, which swings back, however, through readjustments, to the normal for the general area.

At the time of the discovery of North America, the general range of countless numbers of large game animals, such as buffalo and elk, was measured by nearly the full width of the continent, and innumerable antelope roamed the western grassland plains from Alberta and Saskatchewan to northern Mexico. Deer were plentiful, while moose and caribou, mountain sheep, and bears in large numbers occupied vast areas from which they have now disappeared. The larger carnivores, mountain lions, lynxes, bobcats, wolves, and on the western plains, coyotes, followed and preyed upon the game herds, and the surplus afforded food for great numbers. Primitive human tribes armed with bows and arrows existed largely as so many predatory animals, with comparatively slight effect, except locally, on the general balance at that time prevailing.

But with the occupation of the continent by Europeans bearing firearms, clearing the forests, and settling permanently throughout its extent, the balance of nature has been violently overturned, never to be reestablished; and the violence of the overturn has been usually in direct ratio to the density of this new human population. From the early settlements, largely near the Atlantic coast, intensive occupation of the country has spread gradually westward until it now embraces nearly all of the more favorable sections as far as the Pacific littoral. The profound changes brought about by man in his progressive encroachment on the haunts of the larger game, especially along with his too indiscriminate use of modern firearms, have led to the disappearance of most species over much of their former range. Those that remain have been forced to seek the more inaccessible regions, especially the mountains, in the west, fortunately, now largely within the national forests and national parks.

The large predatory mammals—mountain lions and wolves—require for proper sustenance animal food in large quantities. Owing to this compelling need, coupled with abundant power to kill, their preference is for large game, and in their normal diet small animals such as rodents and birds form an inconsiderable part. In order, therefore, to procure accustomed food they have followed the retreating game, or where this is gone and they have been permitted to exist, have turned their attention to the livestock of the settlers. Where numbers of these roving animals are uncontrolled the overturning of the balance that formerly existed tends to concentrate their onslaughts on the already numerically reduced game, threatening its extermination; and unrestrained attacks on livestock have rendered some sections wholly unremunerative as grazing areas.

Since civilized man has hopelessly overturned the balance by creating artificial conditions and contacts throughout his sphere of influence, practical considerations demand that he assume effective control of wild life everywhere. This means that the rodents and other animals, as well as the carnivores, must be checked wherever they become too numerous or too injurious to human interests.

A look backward to the tremendous changes that have taken place during the past century brings to us a realization of the complexity of modern civilization, with the balance of nature, never perfect, now completely overthrown; a look forward to the progressive expansion of the future reveals greater general economic pressure, more insistent demands for the exploitation of unutilized natural resources, and an over-increasing struggle for food. Large predatory mammals, destructive to livestock and to game, no longer have a place in our advancing civilization. To advocate their protection in areas occupied by the homes of civilized man and his domestic animals is to invite being discredited as practical conservationists and to risk through prejudice the defeat of measures which may be vital to the future welfare of the country.

Source: Goldman, Edward A. 1924. "The Predatory Mammal Problem and the Balance of Nature." *Journal of Mammalogy* 6:2.

As the first American to earn a Ph.D. in ecology, Charles Adams holds a distinctive place in the history of science. His career took numerous turns, and he was rarely able to hold a university research position, perhaps due in part to his argumentative nature. Adams worked in natural history museums and strongly supported movements to conserve habitats that could support wild animals. Since so much of the western United States was settled rapidly in the early decades of the twentieth century, he argued in this article that scientists should insist upon the preservation of areas where ecological communities could be studied in something like their earlier natural state.

The Conservation of Predatory Mammals

CHARLES C. ADAMS (1924)

Status of the Problem

In primeval America, before the advent of the Indian, there was a strictly "original" condition, and there has been relatively little since then. With the arrival of the Indian, probably with his dogs, new predators were added to the fauna, and the overturning of the original pre-human relative balance of nature was begun. Wissler (1922, pp. 1–44) has pointed out and mapped the food areas occupied by Indians and the Eskimos of the North with their dogs. Here they preyed upon caribou, seals and salmon. Upon the Great Plains the Indians lived upon the buffalo; while those of the remainder of the continent were mainly dependent upon vegetable foods, supplemented, of course, by a varying amount of fish and game. Thus the predatory habits of the Indians were much more pronounced to the north, and less so to the south.

The Indian population was uneven in its concentration and there were considerable cultural differences among them, so that their influence upon the fauna was probably not at all uniform. Speaking in broad, general terms, the Indian seems to have attained a general balance with the fauna, because his culture was relatively stable, and thus the conditions became quite different from those existing previous to his arrival. Such considerations as these indicate that there was a gross relative balance of nature during the Indian days, and that there were many local variations or minor differences in this balance. The chances are that these were due to local differences in the density of the population, as well as to local geographic influences, the climate, other physical features, vegetation, the animals, and finally to the competing differences among the Indian cultures themselves.

Later the Europeans arrived, slowly at first, but soon in rapidly increasing numbers, and great changes took place. This meant a change of race, other human cultural systems, certain domestic animals, and their hangers-on. The adventurers and trappers invading the North put a new value on the fur-bearing predators and rodents. At once the exploitation of this resource was expanded under the impulse of a keenly competitive system and the old "relative balance of nature" of Pre-Columbian days was soon gone forever. Farther south the farmer-hunter-trapper pioneer pushed settlements and agriculture forward, by clearing away the wilderness and the forest and prairie haunts of wild life, and restricted their habitat. He thus initiated the process of adjustment toward a new balance, suitable for the European. Still farther south gold seekers and propagandists made strenuous efforts to change the Indian culture but they were not so successful as permanent occupants, and the overturning has not been so complete. But everywhere the new transformation has now long been under way.

In the United States our institutions have grown up upon a European foundation, modified to suit the pioneer conditions of a new continent, competing with the native Indian culture in its own environment. There has been thus developed a very complex system of interacting influences in which the native animals have suffered greatly. Our methods of territorial acquisition have given us a Public Domain of great area, which it has been our policy to dispose of as rapidly as possible and, it must be said, often very foolishly indeed.

As long as there were great areas of the bountiful wilderness and of the Public Domain, wild life thrived and was given little concern except for hunting and trapping. This was the adolescent period when we spent lavishly our public inheritance of land and other resources, without much thought for the future, and without careful planning or policy. We have now largely disposed of this great bounty and have begun to feel the pinch of need, and are beginning to come to our senses. With the appropriation of the land, there later developed a great industrialization of the country, the beginning of another era that will involve a new balance between the urban and the rural problem. Many of us have grown up seeing these changes but without an appreciation of their significance and meaning. The best land is now fully occupied, and as well, considerable areas of second class land, and already the abandoned and waste land shows the reality of a new era. This era is now upon us, as has been shown by the public recognition of the conservation movement which was made a national problem by Roosevelt and Pinchot.

Bearing in mind these ideas we are in a position to see how the conservation of wild life—the fish, game and fur-bearing animals—has become enmeshed in this complicated network of economic and social life of our people. We have reached a stage where our naturalists must take a long look ahead, must begin to orient their wild life problems in our general economic and social system, conduct research upon those problems which will aid us in their solution, and formulate definite ideals and policies which will guide naturalists, conservationists and statesmen into proper channels. The solution is not solely a problem for the naturalist, or merely a specialist, because it must be correlated with the activities of progressive students in all the allied lines of economic, social, and political work, so that his policies may give and receive aid, and be properly related to the best general policies.

The preceding account sketches the background of the tragic, losing battle of our wild life, and shows something of the numerous and powerful stresses which have enveloped it. The Indians and the Europeans and their cultural systems have exerted immense, almost crushing, pressure upon predacious animals, and they have resisted this with all the resources at their command. They have shown a spirit of resistance against such odds that it calls for admiration. Is it not fundamentally the same impulse in the animals that resists a similar one in man? To play the game fairly with these animals we will recognize all the merit possible in such opponents, and the naturalist can admire this, even though he does not as an economist.

We see that much energy has gone into the struggle of these opposing systems, and why it is so appropriate to discuss its relation to the so-called "balance of nature." This conception recognizes that in nature there are many influences or forces interacting. Stresses are being exerted, and these initiate the process of adjustment to the pressure. The continuation of pressure will in time tend toward balancing, and finally a state of relative equilibrium will be developed by the balancing of all stresses and pressures.

We may imagine these balances of almost every conceivable magnitude, from those involving any single animal function, the animal as a whole, the species, and as applicable to all sizes of animal habitats or environments. All of these exhibit stages in the process of adjustment to stresses or pressures, or that of a relative equilibrium, caused by the balancing of pressure. Thus, for example, relative over-population produces a condition of stress or pressure upon the local population and on the local environment, and this condition continues until the pressure is relieved or balanced; by dispersal of the animals, destruction of competitors, by death of the excess of population, or by other causes.

As was pointed out, the original animal population worked out its own balance; with the arrival of the Indian a new one was begun, and with the arrival of the European still another major one was started. And it is now our problem to learn to use this knowledge to human advantage, and therefore we have the problem of the conservation of wild life, including, of course, the predators.

Source: Adams, Charles C. 1924. "The Conservation of Predatory Mammals." *Journal of Mammology.* 6:2.

*Raymond Lindeman set the context for ecosystem ecology with his essay
on the transformation and movement of energy in communities of
organisms. Reviewers initially rejected the essay excerpted here, and it
was published only after being shepherded through the process by
Lindeman's mentor, G. Evelyn Hutchinson. Lindeman did not live to
see the influence and success of this early work. Ecologists who had
formerly disagreed over concepts like community and food web found in
ecosystem ecology a productive and mutually agreeable conceptual basis.*

The Trophic-Dynamic Aspect of Ecology

RAYMOND L. LINDEMAN (1942)

Trophic Dynamics

Qualitative Food-Cycle Relationships

Although certain aspects of food relations have been known for centuries, many
processes within ecosystems are still very incompletely understood. The basic
process in trophic dynamics is the transfer of energy from one part of the ecosystem to another. All function, and indeed all life, within an ecosystem depends
upon the utilization of an external source of energy, solar radiation. A portion of
this incident energy is transformed by the process of photosynthesis into the
structure of living organisms. In the language of community economics introduced by Thienemann ('26), autotrophic plants are *producer* organisms, employ-
ing the energy obtained by photosynthesis to synthesize complex organic substances from simple inorganic substances. Although plants again release a
portion of this potential energy in catabolic processes, a great surplus of organic
substance is accumulated. Animals and heterotrophic plants, as *consumer*
organisms, feed upon this surplus of potential energy, oxidizing a considerable
portion of the consumed substance to release kinetic energy for metabolism, but
transforming the remainder into the complex chemical substances of their own
bodies. Following death, every organism is a potential source of energy for
saprophagous organisms (feeding directly on dead tissues), which again may act
as energy sources for successive categories of consumers. Heterotrophic bacteria and fungi, representing the most important saprophagous consumption of
energy, may be conveniently differentiated from animal consumers as specialized *decomposers*[1] of organic substance. Waksman ('41) has suggested that certain of these bacteria be further differentiated as *transformers* of organic and
inorganic compounds. The combined action of animal consumers and bacterial
decomposers tends to dissipate the potential energy of organic substances, again
transforming them to the inorganic state. From this inorganic state the

autotrophic plants may utilize the dissolved nutrients once more in resynthesiz-ing complex organic substance thus completing the food cycle.

The careful study of food cycles reveals an intricate pattern of trophic predilections and substitutions underlain by certain basic dependencies; food-cycle diagrams attempt to portray these underlying relationships. In general, predators are less specialized in food habits than are their prey. The ecological importance of this statement seems to have been first recognized by Elton ('27), who discussed its effect on the survival of prey species when predators are numerous and its effect in enabling predators to survive when their usual prey are only periodically abundant. This ability on the part of predators, which tends to make the higher trophic levels of a food cycle less discrete than the lower, increases the difficulties of analyzing the energy relationships in this portion of the food cycle, and may also tend to "shorten the food-chain."

Fundamental food-cycle variations in different ecosystems may be observed by comparing lacustrine and terrestrial cycles. Although dissolved nutrients in the lake water and in the ooze correspond directly with those in the soil, the autotrophic producers differ considerably in form. Lacustrine producers include macrophytic pondweeds, in which massive supporting tissues are at a minimum, and microphytic phytoplankters, which in larger lakes definitely dom-inate the production of organic substance. Terrestrial producers are predomi-nantly multicellular plants containing much cellulose and lignin in various types of supporting tissues. Terrestrial herbivores, belonging to a great number of spe-cialized food groups, act as *primary consumers* (sensu Jacot, '40) of organic substance; these groups correspond to the "browsers" of aquatic ecosystems. Terrestrial predators may be classified as more remote (secondary, tertiary, qua-ternary, etc.) consumers, according to whether they prey upon herbivores or upon other predators; these correspond roughly to the benthic predators and swimming predators, respectively, of a lake. Bacterial and fungal decomposers in terrestrial systems are concentrated in the humus layer of the soil; in lakes, where the "soil" is overlain by water, decomposition takes place both in the water, as organic particles slowly settle, and in the benthic "soil." Nutrient salts are thus freed to be reutilized by the autotrophic plants of both ecosystems.

The striking absence of terrestrial "life-forms" analogous to plankters[2] indicates that the terrestrial food cycle is essentially "mono-cyclic" with macro-phytic producers, while the lacustrine cycle, with two "life-forms" of producers, may be considered as "bicyclic." The marine cycle, in which plankters are the only producers of any consequence, may be considered as "mono-cyclic" with microphytic producers. The relative absence of massive supporting tissues in plankters and the very rapid completion of their life cycle exert a great influence on the differential productivities of terrestrial and aquatic systems. The general convexity of terrestrial systems as contrasted with the concavity of aquatic sub-strata results in striking trophic and successional differences, which will be discussed in a later section.

[1] Thienemann ('26) proposed the term *reducers* for the heterotrophic bacteria and fungi, but this term suggests that decomposition is produced solely by chemical reduction rather than oxidation, which is certainly not the case. The term *decomposers* is suggested as being more appropriate.

[2] Francé ('13) developed the concept of the *edaphon*, in which the soil microbiota was represented as the terrestrial equivalent of aquatic plankton. This concept appears to have a number of adherents in this country. The author feels that this analogy is misleading, as the edaphon, which has almost no producers, represents only a dependent side-chain of the terrestrial cycle, and is much more comparable to the lacustrine microbenthos than to the plankton.

Source: Lindeman, Raymond L. 1942. "The Trophic-Dynamic Aspect of Ecology," *Ecology*. Vol. 23, No. 4.

Writing this essay near the end of his career, Gifford Pinchot reflected on the dramatic changes that took place in scientific forestry when he was Chief of the U.S. Forest Service. He noted that reserving large tracts of forest land had made it possible for the government to manage resources for future generations. Pinchot's account simplifies some of the struggles over management that involved conflict among settlers, ranchers, wildlife enthusiasts, and political interests from different parts of the country. He nevertheless puts the importance of natural resources on a firm footing, writing in the post-World War II era of expansion.

Breaking New Ground

GIFFORD PINCHOT (1947)

56. A New Science–Forestry

Forty years ago [around 1905] there was in America no science of Forestry, in the true sense of the word. We knew next to nothing about our forest trees, our forest types, their life histories, their enemies, and their friends. Today, while still far from perfection, we have advanced to a fair knowledge of the life of the forest.

Before Forestry came to America, the botanical knowledge of American trees was far advanced, and that was true also of forage plants on the Western ranges. But the practical knowledge of how to use forest and range without destroying them lagged far behind.

With the transfer the Forest Service became responsible for maintaining and even increasing the permanent production of forest and forest range over tens of millions of acres. For that great task, we needed to know more about how different kinds of trees should be marked for cutting, how best to lumber and dispose of the slash, and especially how to ensure natural reproduction. We needed to know also how the range could best be protected and maintained. And we were short on the underlying facts.

To secure these facts, we brought into the Service skilled men—men able to handle our going problems on the ground, and other men competent to carry on the research essential to really good management.

Our previous studies of commercial trees, and what we knew about Forestry in general, had equipped the Service fairly well to deal with the immediate problems of forest management on the National Forests. That knowledge, fortified by practical experience and common sense, got us by without too many glaring mistakes. On the whole, however, it was still too little and too narrow when compared with the vastness of the field and the complexity of the problems.

After the transfer came storm and stress. The great special interests of the West, whose monopolies we threatened, threw against us their lobbyists and their satellite press. Moreover, our men in the field had to meet the old and ingrained habit of exploitation of the Western pioneer. "This is our country. We drove the Indians out. Now you let us alone." It was to take some little time before the people of the West were won over to the sound new policies of the Service. And the only men who could do it were men of action, who could make quick decisions and yet not stumble too much. There was no time for lengthy observation or the search for precise forest facts.

In addition, the Service was growing, and growing fast. One expansion followed close on the heels of another. New technical problems, new executive tasks had to be taken in our stride. Our best men had to be diverted to boundary work, to recruiting new fieldmen, to working out new policies in timber sales and grazing, and to many other problems of our vast new domain. The respect which the Forest Service commands today from the West bears high testimony to the character, courage, and intelligence of these creators of a new forest administration.

The egg of technical Forestry had been laid, but the chick had hardly more than broken out of its shell. Nevertheless, it had to be converted to the practical uses of forest management. And we were lucky that this was so, that no rigid set of rules was imposed upon our fieldmen from the top. Our technical knowledge was born of necessity, and it grew as the need developed. As everywhere in the Forest Service, we gave good men their heads, and the results were good.

In those days the research men were not on their own, but were attached to the Forest Supervisors, to help with technical problems. Their presence was often resented on the ground that they were Eastern tenderfeet (which had nothing to do with the case), but more commonly because of their persistent and sometimes unreasonable habit of asking embarrassing professional questions.

In those early days we usually had to split the difference between what was best for the forest and what was practical in logging. The research men and the practical men had to get together on some golden mean before the marking rules could stand the test of good Forestry and good lumbering. And so with slash disposal, range improvement, and many other technical matters. That research in the Forest Service today gives sound and practical results is due to this constant struggle between what is ideally good and what is practically possible.

Since a forest crop takes longer to mature than any other, there is greater need in Forestry for long and carefully recorded observations than in any other field of agriculture. Such observations do not happen by accident. Even before the transfer there was a little hall room in the Atlantic Building where technical forest facts and statistics were assembled. We called it Compilation. Here was the first cradle and treasure house of forest research in America.

The Conservation of natural resources is the key to the future. It is the key to the safety and prosperity of the American people, and all the people of the world, for all time to come. The very existence of our Nation, and of all the rest, depends on conserving the resources which are the foundations of its life. That is why Conservation is the greatest material question of all.

Moreover, Conservation is a foundation of permanent peace among the nations, and the most important foundation of all. But more of that in another place.

It is not easy for us moderns to realize our dependence on the earth. As civilization progresses, as cities grow, as the mechanical aids to human life increase, we are more and more removed from the raw materials of human existence, and we forget more easily that natural resources must be about us from our infancy or we cannot live at all.

What do you eat, morning, noon, and night? Natural resources, transformed and processed for your use. What do you wear, day in and day out—your coat, your hat, your shoes, your watch, the penny in your pocket, the filling in your tooth? Natural resources changed and adapted to your necessity.

What do you work with, no matter what your work may be? What are the desk you sit at, the book you read, the shovel you dig with, the machine you operate, the car you drive, and the light you see by when the sunlight fails? Natural resources in one form or another.

What do you live in and work in, but in natural resources made into dwellings and shops and offices? Wood, iron, rock, clay, sand, in a thousand different shapes, but always natural resources. What are the living you earn, the medicine you take, the movie you watch, but things derived from nature?

What are railroads and good roads, ocean liners and birch canoes, cities and summer camps, but natural resources in other shapes?

What does agriculture produce? Natural resources. What does industry manufacture? What does commerce deal in? What is science concerned with? Natural resources.

What is your own body but natural resources constantly renewed—your body, which would cease to be yours to command if the natural resources which keep it in health were cut off for so short a time as 1 or 2 per cent of a single year?

There are just two things on this material earth—people and natural resources.

From all of which I hope you have gathered, if you did not realize it before, that a constant and sufficient supply of natural resources is the basic human problem.

Source: Pinchot, Gifford. 1947. *Breaking New Ground.* Copyright ©1947 Estate of Gifford Pinchot, renewed 1974 by Gifford B. Pinchot. Reproduced by permission of Island Press, Washington, D.C.

Aldo Leopold established himself as the founder of game management, publishing a textbook by that title in 1933 and teaching in the nation's first department of wildlife management at the University of Wisconsin. His research advanced theoretical and practical understanding of the requirements of wildlife. At the same time, Leopold became involved in local and national controversies over the implications of his theoretical and practical work. He spent many weekends with his family living out of a shack near the Wisconsin River, and wrote classic essays, including the one here, on the changing seasons and natural features of wilderness, which were collected in A Sand County Almanac *after his death.*

A Sand County Almanac: With Essays on Conservation from Round River

ALDO LEOPOLD (1949)

Wilderness for Science

The most important characteristic of an organism is that capacity for internal self-renewal known as health.

There are two organisms whose processes of self-renewal have been subjected to human interference and control. One of these is man himself (medicine and public health). The other is land (agriculture and conservation).

The effort to control the health of land has not been very successful. It is now generally understood that when soil loses fertility, or washes away faster than it forms, and when water systems exhibit abnormal floods and shortages, the land is sick.

Other derangements are known as facts, but are not yet thought of as symptoms of land sickness. The disappearance of plants and animal species without visible cause, despite efforts to protect them, and the irruption of others as pests despite efforts to control them, must, in the absence of simpler explanations, be regarded as symptoms of sickness in the land organism. Both are occurring too frequently to be dismissed as normal evolutionary events.

The status of thought on these ailments of the land is reflected in the fact that our treatments for them are still prevailingly local. Thus when a soil loses fertility we pour on fertilizer, or at best alter its tame flora and fauna, without considering the fact that its wild flora and fauna, which built the soil to begin with, may likewise be important to its maintenance. It was recently discovered, for example, that good tobacco crops depend, for some unknown reason, on the preconditioning of the soil by wild rag-weed. It does not occur to us that such unexpected chains of dependency may have wide prevalence in nature.

When prairie dogs, ground squirrels, or mice increase to pest levels we poison them, but we do not look beyond the animal to find the cause of the irruption. We assume that animal troubles must have animal causes. The latest scientific evidence points to derangements of the *plant* community as the real seat of rodent irruptions, but few explorations of this clue are being made.

Many forest plantations are producing one-log or two-log trees on soil which originally grew three-log and four-log trees. Why? Thinking foresters know that the cause probably lies not in the tree, but in the micro-flora of the soil, and that it may take more years to restore the soil flora than it took to destroy it.

Many conservation treatments are obviously superficial. Flood-control dams have no relation to the cause of floods. Check dams and terraces do not touch the cause of erosion. Refuges and hatcheries to maintain the supply of game and fish do not explain why the supply fails to maintain itself.

In general, the trend of the evidence indicates that in land, just as in the human body, the symptoms may lie in one organ and the cause in another. The practices we now call conservation are, to a large extent, local alleviations of biotic pain. They are necessary, but they must not be confused with cures. The art of land doctoring is being practiced with vigor, but the science of land health is yet to be born.

A science of land health needs, first of all, a base datum of normality, a picture of how healthy land maintains itself as an organism.

We have two available norms. One is found where land physiology remains largely normal despite centuries of human occupation. I know of only one such place: northeastern Europe. It is not likely that we shall fail to study it.

The other and most perfect norm is wilderness. Paleontology offers abundant evidence that wilderness maintained itself for immensely long periods; that its component species were rarely lost, neither did they get out of hand; that weather and water built soil as fast or faster than it was carried away. Wilderness, then, assumes unexpected importance as a laboratory for the study of land-health.

One cannot study the physiology of Montana in the Amazon; each biotic province needs its own wilderness for comparative studies of used and unused land. It is of course too late to salvage more than a lopsided system of wilderness study areas, and most of these remnants are far too small to retain their normality in all respects. Even the National Parks, which run up to a million acres each in size, have not been large enough to retain their natural predators, or to exclude animal diseases carried by livestock. Thus the Yellowstone has lost its wolves and cougars, with the result that elk are ruining the flora, particularly on the winter range. At the same time the grizzly bear and the mountain sheep are shrinking, the latter by reason of disease.

While even the largest wilderness areas become partially deranged, it required only a few wild acres for J. E. Weaver to discover why the prairie flora is more drought-resistant than the agronomic flora which has supplanted it. Weaver found that the prairie species practice 'team work' underground by distributing their root-systems to cover all levels, whereas the species comprising the agronomic rotation overdraw one level and neglect another, thus building up cumulative deficits. An important agronomic principle emerged from Weaver's researches.

Again, it required only a few wild acres for Togrediak to discover why pines on old fields never achieve the size or wind-firmness of pines on uncleared forest soils. In the latter case, the roots follow old root channels, and thus strike deeper.

In many cases we literally do not know how good a performance to expect of healthy land unless we have a wild area for comparison with sick ones. Thus most of the early travelers in the Southwest describe the mountain rivers as originally clear, but a doubt remains, for they may, by accident, have seen them at favorable seasons. Erosion engineers had no base datum until it was discovered that exactly similar rivers in the Sierra Madre of Chihuahua, never grazed or used for fear of Indians, show at their worst a milky hue, not too cloudy for a trout fly.

Moss grows to the water's edge on their banks. Most of the corresponding rivers in Arizona and New Mexico are ribbons of boulders, mossless, soil-less, and all but treeless. The preservation and study of the Sierra Madre wilderness by an international experiment station, as a norm for the cure of sick land on both sides of the border, would be a good-neighbor enterprise well worthy of consideration.

In short all available wild areas, large or small, are likely to have value as norms for land science. Recreation is not their only, or even their principal, utility.

Source: Leopold, Aldo. 1949. *A Sand County Almanac and Sketches Here and There.* Copyright 1949, 1953, 1966, renewed 1977, 1981 by Oxford University Press, Inc. Used by permission of Oxford University Press, Inc.

Edward O. Wilson and Daniel S. Simberloff undertook a massive ecological experiment in the Florida Keys when they catalogued the species living on a series of small islands before exterminating the islands with pesticides. They wanted to see how quickly the islands would become inhabited once again, and which species would return first. Aside from the obvious logistical obstacles of such an experiment, Wilson and Simberloff combined ambitious quantitative methods with careful field observations in this pioneering work as illustrated in this excerpt. Ecologists have rarely equaled the scale and significance of this experiment, but it has served as a productive model for studies that integrate ecosystem concepts with comparative natural history.

Experimental Defaunation of Islands

Edward O. Wilson and Daniel S. Simberloff (1969)

The Experimental Islands

Along the Overseas Highway of southern Florida (U. S. Route 1) there exist a vast array of small, approximately circular islands situated in shallow bay water. Together, they constitute a potential natural "laboratory" for experimental biogeography. The islands with diameter 10–20 m are 5–10 m tall and usually consist entirely of one to several red mangrove trees (*Rhizophora mangle*), with rarely a small black mangrove bush (*Avicennia nitida*). Occasionally, in areas of weak tide, the trees are surrounded by small areas of supratidal mud and sand. Such land is only intermittently supratidal, since it is completely flooded during prolonged winds of 10 knots or more. Individual islands may differ in several minor respects

TABLE 1. PARAMETERS OF EXPERIMENTAL ISLANDS

	Island name	Diameter (m)	Distance from nearest source (m)	Initial no. arthropod species	Location
Series 1	E1	11	533	25	off Squirrel Key
	E3	12	172	31	in Rattlesnake Lumps
	ST2	11	154	29	off Saddlebunch Key
	E2	12	2	43	off Snipe Keys
	E6[a]	15[a]	73[a]	30[a]	in Johnston Key Mangroves
	E7	25	15	29	near Manstee Creek
Series 2	E8	18	1.188[b]	29	between Calusa Keys and Bob Allen Keys
	E9	18	379	41	between Calusa Keys and Bob Allen Keys
	E10[a]	11[a]	37[a]	20[a]	near Bottle Key

a Control Island
b "Stepping Stone"

such as the number of large trunks and amount of dead bark they contain, but on the whole they are remarkably similar to each other in physical appearance.

A fruitful defaunation experiment with these islands requires the following conditions:

i) that there be enough of the islands for replication and variation in distance to the nearest source area;

ii) that the animal diversity be sufficient to allow statistical treatment, and the organisms be physically large enough to insure recording inconspicuous forms as well as conspicuous ones;

iii) that the extremely small size of the islands compensates for their relative nearness to source areas and therefore produces a distance effect.

During an exploratory trip to the Florida Keys in 1966, we found that conveniently accessible small islands were numerous at most distances up to 200 m from the nearest large islands. Beyond 200 m they were rare, although somewhat larger islands (diameter 50 m and greater) were plentiful. A very few small islands suitable for our purposes were located 0.2–1.4 km from the nearest source area.

A set of such islands were chosen for experimentation and their faunas carefully surveyed. Since the breeding fauna of islands this size consists almost entirely of species of insects and spiders, we made a reference collection of land arthropods of the Florida Keys with the assistance of Robert Silberglied. This collection enabled us to identify the arthropods on the experimental islands, and it provided a measure of the size and composition of the species pool of the presumed source area, the entire Florida Keys.

Animal diversity on the small mangrove islands proved to be adequate for statistical purposes. About 75 insect species (of an estimated 500 that inhabit mangrove swamps and an estimated total of 4,000 that inhabit all the Keys) commonly live on these small islands. There are also 15 species of spiders (of a total Keys complement of perhaps 125 species) and a few scorpions, pseudoscorpions, centipedes, millipedes, and arboreal isopods. At any given moment 20–40 species of insects and 2–10 species of spiders exist on each of the islands.

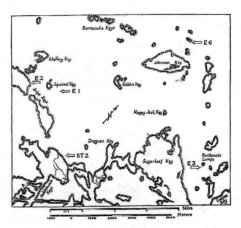

FIG. 3. Map showing the location of the experimental island of series 1.

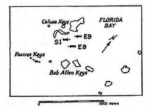

FIG. 4. Map showing the location of the experimental island of series 2.

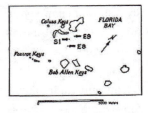

FIG. 5. Map showing the location of the original test island E7.

Many birds visit the islands either to roost or to forage, but very few nest there. The only birds that bred on our islands during the course of the experiment (1966–68) were one or two pairs each of the Green Heron, White-crowned Pigeon, and Gray Kingbird. Snakes (mostly water snakes of the genus *Natrix*) occasionally swim to small mangrove islands, and raccoons visit islands located on shallow mud banks. These vertebrates were not included in the censuses.

On the basis of their distance and direction from the nearest source area, their size, and their accessibility, the islands listed in Table 1 were chosen for defaunation. Two islands (E6 and E10), one each in the vicinity of the two groups of experimental islands, were selected to serve as control islands; and they were censused at the same time as the experimental islands. In order to ascertain whether any long range change had occurred in the general fauna of small mangrove islands in Florida Bay during the course of the experiment, a second census of the control islands was made at the close of the experiment.

The two series of experimental islands form two widely separated groups. Series 1 lies within the Great White Heron National Wildlife Refuge, in a region north of Sugarloaf Key (Fig. 3). Series 2 is in the Everglades National Park near the Calusa Keys north of Islamorada (Fig. 4). E7, which was chosen primarily to test a method of defaunation, is located in Barnes Sound south of the Glades Canal (Fig. 5).

The islands were also selected to show as much variation as possible in the degree of isolation from the immigrant sources. The coefficient of correlation for island series 1 between "distance from nearest source" and "number of arthropod species" is –0.83, indicating the operation in the original faunas of the distance effect as defined by MacArthur and Wilson (1967). Qualitative differences in faunas of increasingly distant islands were even more striking than quantitative ones. For many groups—especially ants and spiders—the distance effect is so regular that one can guess not only the species number but also the approximate species composition. For example, centipedes and pseudoscorpions were found almost exclusively on islands of less than 200 m distance from the nearest large island.

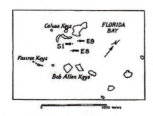

FIG. 6. Map showing E1 and surrounding region as it appeared in 1851–57.

Islands of this minute size can be remarkably long lived. All *Rhizophora mangle* trees, and particularly those on low, small islands, have numerous aerial roots: both brace roots growing out from the main stem and prop roots growing down from branches. The latter can emerge from as high as 4 m above the water (LaRue and Muzik 1954) and in time can become as thick as the main stem. *Rhizophora* trees reproduce viviparously, their seeds

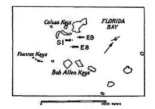

FIG. 7. The same region shown in Figure 6 as it appeared in 1964.

growing on the tree until they become large, long, pointed seedlings. When they drop, frequently from the upper canopy, some evidently plant themselves where they fall (Davis 1940). Mangrove swamps are often quite thick, and on small islands there may be several trees. Stems, thickened prop roots, and brace roots intermingle and reproduce themselves so that it is impossible to distinguish how many trees are present, which are which, which is the initial one (even if it is still present), and which are the main stems.

Source: Wilson, Edward O. and Daniel S. Simberloff. 1969. "Experimental Defaunation of Islands." *Ecology* Vol. 50, No. 2.

Index

About the Author

Chris Young lives and works in Milwaukee, Wisconsin. He is an assistant professor of biology at Alverno College and holds a Ph.D. in the History of Science and Technology from the University of Minnesota. His undergraduate degree, in biology, is from Hamline University in St. Paul, Minnesota. As the author of *In the Absence of Predators: Conservation and Controversy on the Kaibab Plateau* (Lincoln: University of Nebraska, 2002), he has a strong interest in wildlife conservation, scientific controversy, and public understanding of science. At Alverno, he teaches introductory biology, environmental studies, evolution, and seminars on science and global concerns in contemporary society.